Time Mass and the Universe

P.J.Tomlin

AuthorHouse™
1663 Liberty Drive, Suite 200
Bloomington, IN 47403
www.authorhouse.com
Phone: 1-800-839-8640

First published by AuthorHouse 8/1/2008

ISBN: 978-1-4343-3579-1 (sc)
ISBN: 978-1-4343-3580-7 (hc)

Printed in the United States of America
Bloomington, Indiana

This book is printed on acid-free paper.

Salutation and Dedication

This book is dedicated to the memory of Copernicus, Galileo, Newton, Einstein and all those who have dared think beyond the conventional wisdom of their times

Acknowledgements and thanks

The author wishes to express his thanks to Dr Michael Haslam for his encouragement, as well as advice and help in untangling my convoluted sentences and ensuring that all ambiguities were kept to a minimum

Tribute

The author would like to pay tribute to the late Professor Sir Fred Hoyle whose radio lectures fifty years ago first deeply inspired my interest in astronomy, to Professor Stephen Hawking whose writing rekindled that interest and whose lucidity in explanation I envy. Lastly the author pays tribute to Professor S. Perlmutter and his colleagues whose work on distant supernova provided the proof necessary of my theoretical calculations about the Hubble constant—which work was to be the foundation of all the insights described in this book

The author would also like to pay tribute to Archimedes for coining the greatest word in the scientific lexicography, the Hallelujah of Science. That word is Eureka!

Contents

List of Diagrams

Chapter 1

The Prologue

Many cosmologies, like Kipling's Just So Stories, are wonderful constructs for children. As we mature so we learn more. This book is part of that process. This book is intended for the intelligent general reader of science who likes to keep abreast with modern ideas. But it contains new interpretations of previous data, proofs and novel insights that should, it is hoped, be of interest to professional astronomers, astrophysicists and nuclear physicists as well as those interested in relativity theory, or physics in general. Recently the ten greatest problems in modern science were listed by a well known and respected general science journal. Two of those problems related to astronomy. One was the question of dark matter and the other was the anomalous velocity of the Pioneer space probes. More recently still, the same journal pronounced that the greatest mystery of them all was the question of dark energy. In this book a solution to all three is described. What was very unexpected was that the three problems were found to share the same solution.

Among this new information is an understanding of why the Hubble constant generated so much argument. The new information also resolves some astronomical paradoxes, such as why the globular clusters can appear older than the age of the galaxy, and why galaxies can get to be more than ten billion light years away and yet we are able to see their starlight. It identifies the fuel source of gravitational energy, and also the fuel source and controlling mechanism of the expansion of the universe. It identifies and quantifies a fifth fundamental force of nature. A curious relationship between mass and time was uncovered. This relationship was found to underlie our very existence through its effect in the sun. Above all else the book tackles the behaviour of time. All this has been derived from only current published astronomical data and the standard proven laws of physics together with Euclidean geometry. It resulted from looking at some of that data from a different perspective.

In fact this book arose from trying to plot out quantitatively the growth of the universe from use of the Hubble constant, that is the ratio between velocity and distance of any galaxy. But it all did not make sense. There were several obvious paradoxes. This led to questioning every component of the Hubble argument. This in turn led to one simple unanswered question. Has the duration or period of the second remained unchanged since the beginning of time? What was the proof? But whilst working on the possible consequences if the period of the second was not invariant, new data emerged from NASA which provided the additional proof. In particular it was the data from the very distant supernovas that was so helpful.

Perhaps the most surprising was a recent statement from NASA about two of its probes. Pioneers 1 and 2. Both were behaving oddly. NASA has been unable to account for this behaviour. In fact this behaviour provided the final clinching proof for the discoveries described in this book. One of these discoveries predicted that particular behaviour to an accuracy that is at the level of the tenth place in decimals. It was a Eureka moment.

Most people remember the story of Archimedes and how when getting into a very full bathtub he spilled the water. Lying back he began thinking of the complaints from his wife and the female staff, and wondering how much water he had spilled. It is obvious from contemporaneous Greek drama that in their society the women were a force to be reckoned with! It dawned on him that he had spilled a volume of water equal to his own volume. This led him to solving the problem of finding if the king's golden crown had been adulterated with lead. The realisation that he had not only solved his immediate problem but that the solution had wide ranging repercussions and could help resolve any number of other related problems, this was his Eureka moment. Everybody has his or her Eureka moments. Sometimes the discovery is unique to science, more often it is a common place discovery, almost banal, and well known to the rest of the world. But to the finder it is a very special moment of finding something, something that is new to him, that has wide-ranging possibilities and is sufficiently startling in its import for the occasion to be remembered for the rest of the finder's life.

I must confess to having had several Eureka moments in my life; that indeed this book is the direct result of two closely related such experiences. But my earliest was at the age of about 6 years. At that age I was a bright child, I could read simple texts. I knew my multiplication tables. In class tests I was usually among the top group if not the top child. One day the teacher set us all a number of simple multiplication problems. I was confident. I knew the tables. I knew the rules of multiplication, units first, then the tens column. Imagine my dismay when my answer sheet was covered with red ink crosses. It was so bad that I was sent to the principal of the school. She was a fierce no nonsense woman who ruled the school with a rod of iron (but in later life I was to realise that she was an inspired teacher who loved children but stood no nonsense from them). She was teaching a class of 12-year-olds. In fear and trepidation I entered clutching my disgraced answer sheet close to my chest. She glanced quickly at it and set me a couple more sums whilst she carried on teaching her class.

This was typical of my answer sheet

23 x	46 x	55 x	29 x
4	6	5	7
812	2436	2525	1463

She paused, and then set her class some work to do on their own. She then spent three minutes explaining the rules about carrying to the next column and where to put the stray number Suddenly the penny dropped, and dropped with a mighty flash of understanding. I had a solution, no, the solution. It had wide range possibilities. I could do everything with numbers. They were my friends. One very happy small boy trotted out of that classroom.

That was my greatest Eureka moment that has remained with me all my life.

Another Eureka moment came when I was learning about equations. You know the sort of thing, if 3 pears and 5 apples equal 20 apples how many apples make one pear. I resolutely refused to acknowledge the answer on the basis that apples do not make pears. The exasperated teacher (it was the same one) said "Never mind about apples and pears, they can be pooh sticks if you want. Do the sums and work out the meaning later. Don't muddle the two." It was that realisation that calculation and meaning are two different things. That was the Eureka moment. If everything could be put into equation form I could solve any problem (I was only 10 at the time, and a trifle naïve but the principle holds true most of the time). All right, later I would have to work out what the meaning was or what the numbers are trying to tell us. Equations are great tell-tales. But equations became my friends.

One of my friends is the equation £1= 960 farthings where a farthing is an obsolete coin showing a very small amount of money. In Shakespeare's time a farthing could buy you a drink. Today such a drink could cost more than a pound. The equation tells us how much money has devalued since Shakespeare's time, more than a thousand fold. Equations tell us lots of secrets, which is connected with what or who with whom. Who is going to disappear, and perhaps why and so on. Dark family secrets can be exposed-- such as Mass and energy are very closely related indeed. No one ever suspected such an incestuous liaison. But another clandestine relationship hitherto no one has suspected is between time and mass. This is exposed in this book Equations are the greatest of gossips, but they are friendly gossips.

But the question of the calculation and the meaning in reality being two separate things has some serious significance. The calculations of the ancient astronomers showed that the planets revolved in epicycles in a giant celestial sphere that in turn revolved around over a static earth. We now know that the interpretations they made on their data were wrong, even though the arithmetic was accurate. Meaning may or may not have any bearing on reality, like equations that result in positing up to 23 dimensions. Reimann geometry is a geometry of curves and where the angles of a triangle add up to more than 180 degrees. It has many equations, all accurate with the mathematical frame, but they not related to reality. A house built on Reimann geometry lines would quickly fall down as the walls would bulge and overhang and be unable to bear the weight of the roof. But general relativity theory is based on this geometry, which is strange. This is discussed further in this book.

Another set of my favourite equations are the gossipy equations that tell tales. One such tells about mother Earth in particular and what her weight is. The equation looks formidable

$$g = GM_1M_2/r^2$$

But stripping down the equation is all revealing. G is the gravitational constant of known value; g is the gravitational acceleration on earth. It has been measured. It has a value of ~10 metres per second per second. M_2 has a value of unity—on earth the acceleration due to gravity is independent of the mass being accelerated, as dropping bricks and plates from the top of the leaning tower of Pisa has shown. The divisor r is the radius of the earth in metres. M_1 is then the mass of Earth in kilograms with all her overweight glory. Equations are gossips, and who does not like a good gossip. They are my friends. And this book contains a number, not too many, but a number of my friends. The mathematical knowledge required to understand them is simple high school mathematics.

Not all Eureka moments are about mathematics or science. Sometimes they relate to other disciplines. One such moment was the realisation that the application of the law by the courts is not about justice or fairness, nor was it about the resolution or disposal of a problem or claim. It is about the precise meaning of words enshrined in this or that law together with an examination for any hidden meanings or assumptions. Nothing else. Just great precision in interpretation. A precision that matches the finest of Vernier scales an engineer or scientist can devise. But from this interpretation can flow a number of consequences. So also a new word in science should involve similar precision as to its meaning. If not it is merely a label covering something poorly understood. It should be quantifiable or fully descriptive. An example is Space-time continuum. Since both space and time are quantifiable, can this continuum be quantified? What special properties does it have? Does it interact with other modalities? Are there any hidden assumptions? How does this space-time continuum recognise when a large mass is approaching, and that the space then must change its shape and become more curved. What is the mechanism by which space changes its shape? This Vernier precision as to the meaning of words should be applicable to solving other problems in science. Alas it is not always so. Ambiguities and assumptions abound to the extent that they may not even be realised that they are assumptions. One such assumption is the constancy of time.

My latest Eureka moment was the realisation that the duration of the second was not and could not be constant throughout the history of time. For a long time I resisted this but everywhere I sought to test this, and also to test the consequences of this concept no inconsistency emerged. That is it does not conflict with the known laws of physics, but it does resolve a number of astronomical problems, such as dark matter and dark energy and indeed the fuel source of gravitational energy and the expansion of the universe. The ramifications of this extend over many fields.

Well Archimedes discovery had saved the king a significant quantity of gold and what king does not regard favourably any subject who saves him gold. Others have not been so fortunate. Galileo's discovery of Jupiter's orbiting moons and his realisation that Copernicus's theory of a heliocentric universe described the same physical system was less well received. It could have cost him his life. New ideas are always unwelcome especially when they challenge established beliefs or dogma. A principle seems to be emerging that the more established the dogma is, the more inviolate it must be. It has withstood the test of time is the claim, as if that is enough. Any challenge therefore must have a battery of proofs before the old dogma can be jettisoned. It took centuries before science jettisoned Aristotle's concept of the four elements, air, fire, water and earth. It took centuries for medicine to abandon the concept of the four humours in the body. Both had apparently withstood the test of time. This concept of a need for a battery of proofs is wholly unscientific. It would take the finding of just one skeleton of a mammal in Precambrian rock to destroy the entire theory of evolution. By the same token the identification of the distances to the Sloan galaxies on opposite sides of the universe should destroy the current theory of the Big

Bang and the age of the universe. Some of the points made in this book have been presented to the Scientific Journals but were not published because they were either "too controversial", or required "additional proofs" because what was proposed was "so radical". A third Journal of a learned society declined to publish as it was not quite what they usually published but asked to keep a copy of the paper for their Society's library. What was noticeable was the absence of any identification of flaws in the mathematics or scientific deductions. All of which is disturbing but will not cost me my life.

Galileo's other Eureka moment, the discovery of pendulum motion from observing a swinging chandelier whilst listening to a dull sermon, has always been a fascinating one. I can think of few greater pleasures than relaxing and letting one's thoughts wander when there is no possibility of interruption whilst some boring sermon is being preached. Pendulum theory led to a significant ramification in the development some of the ideas in this book. And that started with simply wondering how fast would a pendulum swing in a grandfather or long case clock if it were on the moon that orbited around earth at a relativistic speed. That thought occurred whilst attending a rather dull sermon and my mind had drifted to thinking about Galileo's swinging chandelier whilst also wondering if time had expanded because the sermon seemed inordinately long.

This book was written by a medical academic, one used to analysing quantitatively scientific data and trying to extract the meaning or significance of the observations and their resulting equations. It at first sight would appear strange that a person from one discipline should be involved another. In fact the association between astronomy and medicine is a very long and honourable one, going right back to early Egyptian times. This persisted for centuries. Only 300 years ago, in Oxford, all medical students as part of their training had to study astronomy, although by then most of the astrology had disappeared. (see Paracelsus, The Devil's Doctor, 2006, a biography by Philip Ball, Heinemann, London) This interaction between astronomy and medicine still persists. There has been the fairly recent proposal by astronomer the late Sir Fred Hoyle and his colleagues that epidemics on earth are due to viruses generated in space. Things from space can cause disease, e.g., solar radiation and skin cancer is an obvious example, or radiation (cosmic rays?) can induce mutations in DNA and so congenital disease. But the complexity of the chemical composition of virus particles and the problems of the loading dose of the number of virus particles hitting their target before a disease can affect the individual; renders this hypothesis unlikely. But it is an interesting idea.

This book contains a number of equations, not too many and the mathematical knowledge required to understand them is simple high school mathematics. For those less fluent with mathematics it is suggested that the details of the equations are less important than the arrangement. It is the arrangement that betrays the equation secrets. Mathematics is primarily about arrangements and relationships, the relationship between one item with another. The most famous equation of all, $E=mc^2$ shows the relationship between energy and mass and the arrangement shows that this is a simple linear relationship with a rather large number as a multiplier. The large number is indeed very large. It shows that one gram of mass has the energy that equals a whole day's energy output of Britain's largest power station. But more significant is that the arrangement shows that energy and mass are interchangeable. The numerical details in any equation, although quantitatively important, are of less significance than the relationship displayed by the arrangement. But the beauty about equations is that one can juggle with them and uncover new relationships or expose new concepts.

Another aspect of Nature is that for the most part its numerical system is logarithmic, or exponential. It is just our bad luck that mankind adopted the Arabic linear system of numerical notation such as one to ten and then move to the next column. The logarithmic approach using base ten, where numbers are expressed as exponents of ten, such as 10^2 for 100, 10^3 for 1000 or $10^{0.4}$ (or 2.51, a key number dealing with magnitudes of brightness) is so much more user friendly. This logarithmic aspect of Nature can be seen in the growth of populations. It can be seen in the elimination of drugs from the body, the clearance of pollutants from a lake, the pH of anything, the rate of change of velocity, the application of force, the rate of chemical reactions with temperature etc. The famous, or as some would say the infamous, inverse square law is a manifestation of this. Logarithms are very useful and they eliminate all that tedious multiplication and division. Astronomers use this logarithmic approach in their use of magnitudes to define the brightness of stars albeit their logarithmic base is 2.51 not the more usual base 10.

Definitions and symbols

In this book certain symbols and definitions have been used

Sec is the second and is that period of time that a photon travels 300,000 kilometres (actually 2.9979 x 10^5 kilometres but rounded up for mathematical convenience).

c is the velocity of light rounded up to 300,000 km/sec. It is a fundamental constant. Why it should be this value no one knows but it is.

G is the gravitational constant, that relates gravitational force (g lower case) to the masses of two bodies which are r metres apart. It is another fundamental constant. It has a value of 6.67259 x 10^{-11} N m^2 kg^{-2}

g (lower case) is the gravitational force experienced by a body that is at a distance from some large mass. Since force is mass times acceleration if the mass is 1 kilogram then g describes the acceleration of the kilogram The value of g obeys the inverse square law.

v is velocity. But v (italicised) is velocity expressed as a fraction of the velocity of light, v/c, and is a dimensionless number. Throughout this book this has been used to simplify the equations.

t is normal time, and unless otherwise indicated means one second of present earth time.

T is when time has changed, that is the second is different from normal—either expanded or contracted. The ratio T/t is a dimensionless figure indicating how much time has changed relative to normal time. So that if T/t were less than one the second is of shorter duration than our normal earth time second, that is clocks would go faster. If T/t is greater than one then time has slowed, or the second has expanded. It was Einstein who first showed that time would expand if you moved fast enough.

H_O is Hubble's constant describing the rate of recession of a distant galaxy per Megaparsec. Its units are kilometres per second per Megaparsec. Dimensionally it is velocity per unit distance or bizarrely the reciprocal of time, that is it is a frequency

H_T is a new symbol. It is the Hubble constant where the divider of velocity is time, where time is that time taken by light to travel the Megaparsecs used in H_O. It is in units of metres per second per second and, as will be shown in this book, is an another fundamental constant describing a newly discovered fifth force in nature. Like all fundamental constants it is constant in all time frames. That is it has the same value relative to the pace of time of its immediate environment. When viewed from a different time frame, where time is faster or slower, it will appear to have changed.

ly is light year. The author prefers this to the parsec as it makes the relationship between time and distance clearer. It also simplifies a number of equations. It is about ten trillion (10^{13}) kilometres. Related measures are the light sec, or the distance travelled by light in one second (300,000km). Similarly the light hour is that distance traversed by a photon in one hour (the solar system as a radius of approximately ten light hours) and so on.

°K   is temperature in degrees Kelvin. This is the temperature that starts at absolute zero or -273.15 degree Celsius but each change in degree equals that of a change in degree on the centigrade or Celsius scale. Thus 293.15 corresponds to 20° Celsius.

J   is Joule, that is a unit of energy that is available at any time.

W is watt and is the energy used per second, that is Joule per second.

f   is frequency per second, measured in Hertz

λ is wavelength

h is Planck's constant, another fundamental constant involving energy and time.

Parsec is the parallax arc second, and equals 3.26 light years, that is the distance light travels through space when travelling for 3.26 years

One other factor needs to be borne in mind. Great thinkers and discoverers do make mistakes. These mistakes may not be challenged for years because of the magnitude of honour and respect shown to those persons. Nevertheless they are human and are expressing views consistent with the knowledge available at the time. One can think of Aristotle and Galen with their views of the four elements, fire, water etc., or of the four bodily humours. Even Isaac Newton made mistakes as his belief and experiments that it was possible to transmute ordinary metals to gold. In more modern times the errors may be in the detail rather than the principle involved in their discoveries. Edwin Hubble first identified and quantified the expansion of the universe. He grossly over estimated the expansion constant but that does not detract from his greatness in identifying the expansion of the universe and showing that it had a mathematical basis. Even Einstein made mistakes as he acknowledged, although what is now suggested in this book is that his self confessed mistake was not a mistake, rather, that lay elsewhere.

Greatness should not preclude critical examination of the detail of the works of the great. Even the great ones are human and such showing enhances their greatness. We empathise with the human like weaknesses shown by the great ones in cosmologies that were proposed in the past, such as the Greek gods and goddesses such as , Aphrodite, Ares, even Zeus himself, rather more than the perfect Pallas Athena, although she remains our ideal. Equally and for much the same reasons in the Scandinavian cosmology we appreciate and respect Thor, Loki and even Odin, despite or perhaps because of all their shenanigans, rather more than Balder the perfect one. This book therefore is not an attempt at a put down of great men but rather to emphasise their greatness by showing that they were human. The old phrase is "to err is human…"

To paraphrase a very great man, Isaac Newton, who said "I have stood on the shoulders of giants." in this book all I can say is that. I have peeped over the shoulders of very, very great men and what I found was wondrous to behold.

Chapter 2

Problems in Cosmology

The problem of size, the Sloan galaxies and their effect on the Big Bang hypothesis, factors relevant to expansion, the problem with time, the data tools behind quantitative astronomy.

The universe is big. That is a massive understatement. It is vast. Our nearest neighbour is a mere four light years away, that is forty trillion kilometres away. (40×10^{12} kilometres) The furthest galaxy seen so far seen is twelve billion light years away, (16×10^{21} kilometres, sixteen thousand billion, billion kilometres). That very distance creates all sorts of problems with current cosmological theory.

The universe is packed with galaxies. They vary in shape and size. How many? A rough approximation can be obtained but it rests on two assumptions, that our sun, a standard G type star, is typical of the majority of stars in our own Galaxy, that is the Milky Way, and that our Galaxy has average mass. For various esoteric reasons it has been deduced that there are about 10^{80} nucleons (protons and neutrons) in the universe. Avogadro's number 6×10^{23} is the number of elementary units in a Mole, that is the molecular weight, expressed in grams, of any substance. It follows that there are $\sim 3 \times 10^{23}$ protons per gram of hydrogen. (Molecular weight ~ 2) Hydrogen is by far the commonest gas in the universe. The sun has a mass of approximately 2×10^{33} grams. That is it consists of approximately 6×10^{56} nucleons. There are about a billion stars like our sun in the Galaxy. This amounts to about 6×10^{65} nucleons. If there are 10^{80} nucleons in the universe and each galaxy has 10^{65} of them means that there are about 10^{15} galaxies. That is there are about a million billion galaxies, if they were all of the size of our own Galaxy. This corresponds to two hundred thousand galaxies per person on this earth with each galaxy having a billion stars each as big as the sun. These figures are extremely crude but indicate just how big the universe is.

It is also remarkable heavy. The sun has a mass of 2×10^{30} kg. Our galaxy with its billion suns has a mass of 2×10^{39} kg. And if there are 10^{15} galaxies in the observable universe this gives a total mass for the universe of 10^{54} kg. If all this mass were congregated in one place it would create a black hole more than four thousand times bigger than our galaxy. Because of the fact that gravity slows down time this has a significant repercussion on any estimate of the age of the universe. Current cosmology would also have the universe starting as a single very tiny spot, a singularity. But with that mass it would have been the grand daddy of all black holes. But nothing can escape a black hole So how did the universe expand? How did photons, electrons, protons etc., escape gravity's crushing maw? This is dealt with in detail in Chapter 10.

What is also astonishing is with this vast number of galaxies is that they are distributed throughout space remarkably evenly. There are, on a small scale, groups of galaxies, a cluster of a few hundred here or there. A great wall of several hundred thousand, perhaps a million galaxies has been observed. But the numbers in these groupings are small compared with the overall number, of a million billion galaxies.

Recently a most unusual stellar map has been devised by R. Gott and M. Juric of Princeton University (2004). This displays the stars and galaxies (there were 126,000+ galaxies in this map) that would appear directly overhead at the equator during twenty-four hours in a narrow strip that is only a few degrees wide. The scale is logarithmic which results in an apparent congestion of galaxies at the high distances. These very distant galaxies are the Sloan galaxies. The furthest is well in excess of eleven billion light years distance with at least one supernova observed at over ten billion light years distance. The advantage of this display is that it indicates that whichever direction one looks there are galaxies that are over ten billion light years away. This set of observations undermines all current cosmological theories, including the Big Bang.

In broad outline current cosmology has it that from a minute singularity came a huge surge of energy that rapidly expanded. This was the Big Bang. Some of that energy condensed to form electrons and positrons together with protons and antiprotons. These in turn went through a phase of mutual annihilation, the resultant energy released formed more nucleons and the cycle was repeated almost but not quite endlessly. For some unexplained reason there was an asymmetry resulting in an excess of protons over antiprotons with each generation just as it was with electrons and positrons? Eventually a universe dominated by electrons and protons emerged. And thus our universe was born. The electrons and protons eventually formed galaxies.

Meanwhile space was expanding and carried with it the galaxies so that they became scattered throughout space. But there is this oddity that the universe is remarkably isotropic, that is galaxies are distributed very evenly throughout all space. That means the energy from which the galaxies were derived must have been remarkably evenly distributed at the time of the Big Bang. The energy would have had to have moved faster than the present speed of light for an extremely brief period to ensure the uniformity of distribution of energy, and so mass, and eventually a uniform distribution of galaxies. But light is energy. That is the speed of light could not always have been constant. This episode has been called inflation. Initially the phase of inflation was calculated to end when the universe was less than a cubic metre in size. Doubts have grown about this size. This whole process started about 13.5 –13.7 billion years ago. A more detailed analysis of this cosmology is to be found in Chapter ten

Two major ingredients of this cosmology are that the universe is expanding and the age of the universe. Both have serious problems.

The expansion of the universe depends upon the precise meaning of the term universe. Originally to the ancient Greeks, who coined the term, it meant everything that was or could be. They had no concept that there could be things in the heavens that was beyond their vision. With the development of better and better telescopes the universe was shown to be bigger and bigger. The galaxies are moving away from us at an increasing velocity and by implication this means that space is expanding. This is said to account for why we cannot detect the high frequency electromagnetic waves that would have existed when the universe was very hot. The waves have, it is said, been stretched out by the expansion of space to become the low frequencies of the background microwave electromagnetic radiation.

There are many difficulties in this scenario. There has been no explanation for the source of energy that would drive the expansion of space. To accelerate a galaxy takes a very considerable amount of energy. The observation of galaxies colliding with each other belies the possibility of an adhesive link between galaxies and space which is meant to enable space to drag the galaxies apart. Something must be pushing the galaxies. Many of the galaxies have a specific structure that is replicated with others. These are the spirals, barred spirals and many variations of this theme. But the significant thing is that the galaxies are composed of a mixture of masses, which vary from black holes to tenuous clouds of gas. Each component is only linked by the faint pull of gravity over the very long distances within the galaxy. If there were an external force pushing them outwards the spiral galaxies would not be able to maintain their structure. That is each component of the spiral galaxy has just enough energy to push that component along *parri passu* with the other components of the galaxy. Pushing the galaxies, and accelerating them (and acceleration has been observed,) requires a lot of energy. That energy must arise from within the component itself, be it a dust cloud or a massive star. That is the energy must be mass related. One subsequent thought is if this energy arises from within each component then that energy ought to enable the mass to escape from a black hole unless it can do its work within the black hole's event horizon. But by definition nothing can escape a black hole. Time also stops within a black hole. But energy is time related. That is the ability to do work to push emaciated nucleons against gravity is intensely compromised.

Perhaps the most serious objections to the concept of space physically expanding and so stretching electromagnetic waves lie in the physics of radiation. Fundamental to everything is that the velocity of light is constant (relative to the time frame or how fast the clock is ticking). The expansion of space requires a radial expansion. That is the outermost parts would be displaced outwards. Any light travelling in that section of space should be carried outwards. The consequent effect of the expansion of space would be that the speed of light in that rapidly moving block of outer space would be much faster than the speed of light closer towards the centre of the universe. But the famous Michelson and Morley experiments showed the velocity of light was constant and independent of any direction of movement of the observer or space around the observer. It follows that space itself is not physically expanding. The corollary is that there is more space, empty space beyond the observable universe and that the masses, galaxies etc. are merely expanding into that empty space.

To digress briefly. The Michelson and Morley experiment was designed to see if the earth's velocity and direction of movement made any difference to the measurement of the velocity of light. It showed that the velocity was the same in all directions. If one takes the definition of the second as that period whereby a photon would travel 3×10^8 metres then clearly there is no problem. Time is the same whether we face East or West. What clearly is happening is that the velocity of light is constant relative to its time frame? A photon entering earth's time frame would adjust its speed accordingly.

It has been reported that an experiment involving a pair of matched atomic clocks, when placed on two fast moving aircraft, so that one went East and the other travelled Westwards, there was a discrepancy, even though the aircraft maintained the same air speed. That is time was stretched more when one was travelling in one direction that in the opposite direction. Substitute photons for pilots in the two aircraft and the apparent inference is clear. That the expansion of time changes with direction of the velocity. This inference completely undermines all Relativity theory. The Special theory of Relativity critically relies on the equation that states that the expansion of time rests firmly on the velocity irrespective of the direction of that velocity. Hiroshima and Nagasaki are stark evidence of the truth of the Special Relativity theory. That is special relativity produces the famous $E=Mc^2$ and so predicted the power of the atom.

So what is happening with these two atomic clocks? The experiment reported the influence of travelling East or West. This seems to be the key. The earth is revolving so that when flying eastward the destination is moving toward the aircraft. That is the distance through the air that the plane has travelled is different from the ground-based measurement. The elapsed or flight time would be different. Allowing the atomic clocks to stand after landing would eliminate this problem. Yet the clocks showed that they were out of synchronicity. But what also has to be taken into consideration is that when flying Eastwards the air that is on the route to the destination is also moving with the earth's rotation, creating a headwind. More significantly the velocity would be relative to the plane's immediate environment and not the ground. For a plane flying westward the earth's rotation would create the effect of a following wind, which would also alter the speed of the aircraft relative to its spatial environment. From the point of view of the expansion of time relative to the velocity, each plane would have experienced a different velocity and therefore it is not surprising that the atomic clocks registered different times.

Another possible factor is the geography of the two routes. If one route passed over mountains whilst the other was over the sea that the gravitational experience of the two craft would be different. But time is affected by gravity.

Thus it emerges that the velocity of light is the same in all time frames.

Returning to the stretching hypothesis, another objection is if the light consisted of a long train of waves—an hour's burst of light would generate a train of waves over a trillion metres (10^{12}) long. Any expansion during their transit would result in the photons in the leading waves travelling faster than the rearmost as the train elongated. That is the speed of light would not be constant.

Another objection to the stretching hypothesis is the observation that the light spectrum of a type one supernova is constant irrespective of the distance, as judged by brightness, apart from red shift due to the local velocity. The observation of different red shifts of stars on opposite sides of a rotating spiral galaxy, when viewed edge on, also contradicts the expansion of space elongating wavelength hypothesis, since the two stars are essential the same distance from earth. That is red shift is not related to the presumed expansion of space. Such differences can only be determined for relatively nearby galaxies because telescopes cannot resolve individual stars in very distant galaxies. But such differences in the velocity of stars on the edge of galaxies have been used to calculate the Hubble constant and resulted in figures which reasonably match other means of determining the Hubble constant for nearby galaxies. This has profound significance when considering dark matter.

Yet the background microwave radiation does exist but the physics require a different explanation from that of the microwaves being stretched. But this sort of thinking raises serious questions as to the source of the microwave radiation. Of particular concern is that the intensity of the radiation is the same whichever direction one looks. The radiation is approaching earth. This raise questions what has been the journey for those microwaves. Where have they been? The strength of the radiation suggests that irrespective of the direction they are coming from the microwaves have travelled the same distance. If the distances were varied the inverse square law would predict different strengths of the microwave signal according to the direction.

The fact that the waves are coming from different directions implies that if they originated from the Big Bang there must be some kind of reflective surface. Could it be that the Big Bang explanation is wrong? One possible but very tentative proposal appears later in this book.

A simple resolution of some of these difficulties with the present cosmological hypothesis emerges from considering the precise meaning of the term universe. If the meaning is, as was originally stated by the Greeks, simply all that is visible, then the possibility emerges that there is a potential infinity of empty space beyond the visible universe. It also means that the masses (stars and galaxies) that make up the visible universe are moving into that empty space, so increasing the volume of space within the visible universe. The very far and presumed empty outer space could well contain other universes but we can never know, neither now nor in the future. Neither can we know their sizes.

The other key difficulty with the Big Bang hypothesis is the question of how old the universe is. Current projections are that the universe is around 13.7 billions years old. This is based on a back extrapolation of the rate of expansion, the Hubble constant. Using different galaxies this consistently results in around 13.5-13.7 billion years. Yet the Sloan galaxies are more than ten billion light years away. Simple geometry suggests that the sum of the distances, in light years, of two galaxies that lie in opposite directions cannot exceed the age of the universe in years. It would take time for the two galaxies to reach their present positions from the site of the Big Bang, generate stars and supernovas whilst en route, and then shine their light back to us. That light would take time to travel. The Sloan galaxies are in excess of ten billion light years away. This allows only 3.7 billion years at most for those galaxies to reach that position. For many that are beyond ten billion light years distance the available time for the initial photons to travel that far distance and then condense to form galaxies is even less. But nothing normally can travel faster than light.

Inflation is defined as a period where energy did travel faster than light. But if this were the case the inflation would have had to continue until the radius of the universe was over seven billion light years. The maximum radius of the universe then becomes about twenty one billion light years (7 plus 13.7). The volume of the universe thereafter would have increased only nine fold from the end of this mega inflation. The basic gas laws in physics state that expanding a volume to twice its size halves the temperature. Conversely contracting the volume to half size doubles its temperature. When the volume is one-ninth its temperature will be nine times higher than when at normal volume. Given that the mean temperature of the visible universe appears to be around 2.3°K this puts the temperature of universe at the end of the inflation as 20°K. This is far too cold for hydrogen fusion to form helium. By definition the fusion could not have occurred before inflation, whilst had it occurred whilst inflating the gravitational stresses and stretching of the hydrogen clouds would have prevented condensation of the clouds to form galaxies.

There are two time-related paradoxes. The first is the distance paradox. If the Hubble constant is ~50 km/sec/Mega Parsec, and many observations of nearby galaxies confirm this, then if the constant is constant throughout the universe the Sloan galaxies when judged by their velocities are older than the universe. It would have taken those galaxies 19.8 billions years to get as far as 10 billion light years away. If they got as far as twelve billion light years distance it would have taken 21.6 billion years just to get there. Add to this the time for the light to come back to us, another ten billion years, then clearly not only are these galaxies older than the universe but clearly the age of the universe is wrong. Even if one accepts that the Hubble constant is 70 km/sec/Mega Parsec it would have taken 16.7 billion years to reach a distance of 10 billion light years. Add to this the transit time for the light to reach us then clearly there has not been enough time given that assumption that the universe is only approximately 13-14 billion years old.

A similar paradox exists for the globular clusters within our Galaxy. There are over 150 clusters of stars. Each cluster has anything up to a thousand stars all closely networked together within the cluster. For any one cluster all the stars share the same common property of being extremely primitive stars, as judged by their spectral composition. The spectra do vary slightly between different clusters, that is the age of the clusters is somewhat variable but nevertheless they are all very old. These clusters are dotted throughout the Galaxy and their age, judged by their brightness, is more than 13.5 billion years. A recent finding of one isolated star with the same spectral pattern as those in the clusters and whose distance could be ascertained with reasonable accuracy suggested that the brightness of stars within clusters had been under-estimated because of errors in estimating their distance. They are further away than previously thought. Their dimness is the result of distance as well as age, and so they could be within the putative age of the galaxy, but it is a closely run thing. The application of one observation of one star to the many thousands that make up the clusters renders this conclusion as tentative. Not least is the question what is an ancient star doing wandering through the Galaxy on its own, and therefore can it be taken as representative of the stars within the clusters. This then is the globular cluster paradox, that our Galaxy contains stars, which are or could be apparently older than the Galaxy itself.

Another problem is how long does it take for a giant gas cloud to condense to form a galaxy. Galaxies have been observed as far distant as 10+ billion light years away. That is they are older than ten billion years. But the estimates of the distances are based on supernova explosions. Supernova explosions in turn mark the end of life of a star. Giant stars have the shortest life but this is of the order of 1-2 billion years. This leaves only 1-2 billion years for galaxy formation. There simply has not been enough time for this if the age of the universe is 13.7 billion years

Fundamental to all the problems is the assumption that time, the period of the second, has remained the same since time first began. There is no proof that this is true. It is an assumption that has become steeped into the mindset of all cosmologists and yet useful calculations can be made on the basis of this assumption.

The data tools of astronomy

The key components in the study of astronomy are: knowing the distances to the various astronomical objects, knowing their velocity, knowing their mass and composition and knowing their radiative output.

Distance determination.

There is no one single method of determining interstellar and particularly inter galactic distances. Each method has its usefulness over a range of distances and some overlapping provides the continuation for accurate measurement in the distance ladder. Nevertheless for most long distance measurements an error of up to 10% has to be allowed for.

For objects close by and within our Galaxy the parallax measurement is the method of choice. In this method the angle needed to bring the object into sharp focus is measured from two widely spaced positions. If you know the distance between the two positions and the angles then it is a matter of simple geometry or trigonometry to determining the distance. This can be done from earth for very nearby objects such as the sun but for more distant objects the base distance is the diameter of the earth's planetary orbit, a distance of nearly 17 light minutes. A parallax angle of 1 second of a degree, or arc second, corresponds to a distance of 3.26 light years and is the parsec. The difficulty of measuring an angle less than a fiftieth of an arc second are such that the method fails at distances a little beyond ten parsecs, about fifty light years. By taking a group of stars that apparently are close together one can achieve a statistical parallel determination which improves the distance measurement by a factor of approximately three fold. The angular resolution is also improved by almost ten fold when making measurements using orbiting satellites when the blurring of the image by air currents within our atmosphere is lost.

The next development was to determine the velocity across our line of sight and then determine what the change in position in arc seconds was after some long period. This is based on the conclusion that groups of stars do rotate around each other. Hence measuring the transverse distance over a period of time is the equivalent of measuring the change in radial distance over this same period of time. This improved the resolution to several thousand light years. Alternative measurements using brightness are also useful if the composition and colour of the object are known. Comparing stars of similar characteristics by their brightness enables distances to be determined if the comparitor's distance is known.

This approach led to the next step up the cosmic distance ladder, using variable stars. The first set was the RR Lyrae. These are short period variables whose cycle of brightness lasts only hours. Triangulation and parallax methods have established their distances as they relate to their brightness and this had led to estimates of distances well in excess of three thousand light years. For comparison our Galaxy has a radius of fifty thousand light years. RR Lyrae have enabled the determination of the distances of a number of the globular clusters Most globular clusters have at least one. They are common in our Galaxy. The RR Lyrae are not particularly bright, in fact too dim to be of use for determining inter galactic distances.

The next class of variables is the Cepheid variables. These are pulsating stars, but the period of pulsation of a Cepheid is directly related to its brightness. The longer the period the brighter the star. The cycle of their pulsations is measured in days and at the end of the cycle they are very bright. Cepheids are comparatively common. The Hipparcos satellite found 220 in a sample of 120,000 stars. A galaxy of a billion stars could well contain a hundred thousand Cepheids. They represent the terminal stages of a large but dying star. They are very rich in helium. Heating the helium hot enough causes it to ionise. At a high temperature the ions become doubly ionised in that each ion loses yet another electron. These doubly ionised ions block the radiation of light and heat (in principle similar to a greenhouse gas) and this causes the star to over heat and swell. The surface enlarges and this cools causing the doubly ionised helium to drop to a single ionised state, which is much more transparent to heat and light. The excess heat is radiated off and the whole star shrinks back to normal size which heats up the outer envelope to form the doubly ionised helium and so the cycle is repeated. It was observing a number of Cepheids in our neighbouring Andromeda galaxy that enabled its distance to be determined. But Cepheids are common in our own Galaxy. The

Cepheids could act as standard candles of known brightness. It is then a matter of the inverse square law to determine the distance to another galaxy.

An improvement on this was the realisation that determining the velocities of galaxies at different times could determine the rate of acceleration and so the velocity across the line of sight. Knowing the velocity and the time taken to cross as section of sky enabled the distance to that section be determined. Of interest the position of certain stars was aligned with man made monuments at the times when the monuments were built, e.g., the pyramids. Since then the star's positions have changed. The time taken to achieve this change is known and so the velocity is known. Together these advances enabled the determination of distances in excess of several million light years. But this is pitifully small compared with the size of the visible universe.

The latest development is the use of a type one supernova as a standard candle. This type of supernova is the result of a white dwarf star accruing to itself sufficient hydrogen that it becomes very unstable. At a specific critical mass it explodes as a supernova. It thus releases a set amount of energy as light, the brightness of which fades in a characteristic manner over the next fourteen days. Tracing this enables the peak brightness to be determined, and comparing this with the brightness of a similar supernova at a known distance (ten parsecs) the distance can be calculated. At very long distances, or high magnitudes, that is beyond five billion light years, this procedure of calculation becomes susceptible to substantial error as magnitude or brightness is a logarithmic scale. A small error in the measurement of the magnitude then has a disproportionate effect on the calculation of distance. A standard type 1a supernova has a magnitude of −19.7 when at a distance of ten parsecs. In contrast the sun would have an absolute magnitude of +4.8 if it were at this standard distance of ten parsecs. Magnitude is a negative logarithmic scale with the base of 2.51. A unit increase in magnitude represents a 2.51 fold decrease in brightness. That is the difference in magnitude between the sun and a Type 1a supernova represents a difference of more than five billion times difference in brightness. The light from supernova positioned at ten parsecs or 32.6 light years from earth would be considerably brighter than a full moon despite the source being considerably more distant than many of our neighbouring stars. Thus supernovas are extremely bright. They are detectable at distances in excess of twelve billion light years. They are not common, occurring about once every three hundred years in a galaxy. But such is the number of galaxies that large numbers are detected every year. Supernovas can also occur deep in the heart of a galaxy where they may be a substantial amount of interstellar dust. The Horse Head nebula in our Galaxy is a characteristic example of interstellar dust. The dust in the heart of a galaxy could absorb some of the light and so cause an error in determining the true relative magnitude. A general principle is that supernovas of the same velocity should be at the same distance, so that pooling results from different supernovas of similar velocities gives a better estimate of distance, which can then be applied to any anomalous observation.

Gravitational lensing can also magnify the brightness and so distort the measurement. Fortunately gravitational lens are very uncommon.

Velocity

The Doppler effect. When a moving body emits energy as a wave, e.g., sound energy, then if that body in moving towards you the pitch or frequency rises. Equally if it is moving away from you the pitch falls. The reason is that the wavelength is changing according to the velocity of the object. The change in pitch of an approaching train is an oft-quoted example. The relationship between velocity and the ratio of wavelengths λ is given by the equation

(Equation 2,1) $$\frac{\lambda_1}{\lambda} = \frac{(1+v)^{0.5}}{(1-v)^{0.5}} \text{ or square root}$$

Where λ_1 is the wavelength whilst the emitter is moving at velocity v, (as a fraction of the velocity of light) and λ is the wavelength when the emitter is stationary

A complication of the way this is reported is the use of the z number where z is the increase in wavelength, $\delta\lambda$, divided by the wavelength, thus

(Equation 2,2) $$z = \frac{\delta\lambda}{\lambda}$$

So that $$\frac{\lambda_1}{\lambda} = \frac{\lambda + \delta\lambda}{\lambda} \text{ or } 1+z$$

A z value of 3.5 corresponds to a velocity of 0.9c, where c is the velocity of light, and a z value of 5 corresponds to a velocity of 0.95c. Such velocities have been reported when examining the spectra of distant quasars.

Of some significance to what is presented later in this book, if the time frame changes, that is the emitter is located in a place where time is faster than our present pace of time and the wave train enters a zone of slower time, it merely travels more slowly. Relative frequency and wavelength remain unchanged.

Frequency, the number of times the photon appears to oscillate to create a wave is the result of photon emission from a heated atom. The photon is the result of an electron circling around an atomic nucleus changing its orbital circuit around that nucleus. The force for the photon's energy is supplied by the heating source. If the force applied is low the frequency will be low and this results in a lower frequency of longer wavelength. This is seen in the dull red heat of a moderate heated mass, compared with the white heat of faster frequencies when the mass is heated vigorously. But force has a relationship to the pace of time (actually to the inverse of the square of time). If time is fast relative to our time then the force is fast and so the frequency. That is frequency is also constant in all time frames. When photons pass into a zone of slower velocity, as opposed to slower time, c the velocity of light slows down. But frequency is unchanged, where frequency is relative to a distant observer. Wavelength shortens. That is the waves bunch up. This explains why when viewing colour under water the colour is unchanged, even though the velocity of light in water is much slower than light in air. A slightly different situation occurs when photons enter a zone of time slowing, say around a large mass. Wave bunching must still occur as the incoming train of waves piles in. In addition to being affected by the medium in which it is travelling the velocity of light is always relative to the local pace of time and frequency must adhere to this. Because c the velocity of light is constant in all time frames it follows that wave frequency is constant in all time frames. When considered from a slower time zone the speed of light is apparently increased. A general principle emerges. All things that are time dependent, velocity, acceleration, forces (including gravity) and frequency adhere to the prevailing pace of time.

Radiation emission

All bodies emit radiation. The type of radiation emitted depends upon the rate of energy being applied. When this is extremely intense so that a lot of energy has to be emitted very quickly this results in radiation of shorter wavelength. If the supply rate is lower the result will be infra red radiation or even radio waves. If the rate is very fast it will result in X-ray or even gamma ray radiation. The rate of the radiation energy being emitted will in turn depend upon the rate at which mass is being converted back to energy. This also applies to gravitational energy although the energy packets, gravitons, are not electromagnetic. Changing the direction of movement of fast moving particles also results in intense short wave radiation. This has been attributed to friction but since the particles are charged, charge repulsion should prevent physical friction. In addition most if not all the particles will be travelling at the same velocity. There is no evidence that friction is responsible for radio galaxies.

Light is the principle radiation. It is measured as brightness using an obscure system of magnitude where magnitude is the negative logarithm. It is somewhat like the more familiar pH scale where the larger the pH value the less acidic is the solution. The pH scale means that unit increase in pH value represents a ten fold reduction in acidity (or hydrogen ion concentration). With magnitudes, instead of base ten as used in the pH scale, for brightness determination the base used is 2.51 That is each change in unit value of Magnitude corresponds to a change in brightness of 2.51 fold. Because the magnitude scale is negative an increase in one unit of magnitude represents a diminution of brightness by a factor of 2.51. Excessively bright objects therefore have negative magnitude values. Human eye discrimination can distinguish up to six orders of magnitude or approximately 250 levels of brightness. The best telescopes can see up to 25 magnitudes or have a one billion range of brightness. Brightness follows the inverse square law with distance and it is this, which has led to the concept of using standard candles as a means of estimating distances. It suffers from being vulnerable to anything which could intercept some of that faint light, such as dust in our galaxy or in the emitting galaxy. Because of the nature of the inverse square law and the logarithmic scale for bright objects that are reasonably near, an error of half a magnitude will cause comparatively little error in determining distances. But with very faint objects a long way away a loss of half a magnitude could cause an error of many millions if not billions of light years in the estimation of distance. Magnitude also depends upon how hot the object is. This will also alter the frequency of the emitted light and so some compensation may be necessary. For this reason measurement of brightness for very distant objects is made at more than one wavelength or frequency.

Mass and its composition

The determination of mass requires use of one of the two equations of gravity. The first is the equation defining the gravitational force at any particular distance (r) separating two masses. This is

(Equation 2,3) $$g = \frac{GM_1M_2}{r^2}$$

where the M is the mass of bodies 1 and 2 respectively and G is the gravitational constant. This latter is a fundamental universal constant that is constant in all time frames. That is it has the same value relative to the pace of time, just as c the velocity of light has.

To calculate the mass of the earth, hold a 1 kg mass (M_2) 1 metre from the surface of the earth and measure the gravitational acceleration when allowed to fall (g, which is approximately 10 m/s/s). Then from G (6.672×10^{-11} Nm²/kg²) and knowing the radius of the earth, the denominator r, it is possible to calculate the mass of the earth. This has a value of 5.976×10^{24} kg.

The mass of the sun can be calculated from another equation, which relates the gravitational force (g) required to keep a body in orbit around a central mass, and at a particular velocity and distance. The orbital period of the earth around the sun is known (1year) and the distance (approximately 50 light seconds) is also known and the velocity can therefore be calculated. Once this is determined then the above equation can be used to determine the mass of the sun. It has a value of approximately 2×10^{30} kg. That is the sun is three hundred thousand times more massive than the earth.

Stars vary in size, and lettered according to their size. A G star such as our sun is the commonest. There is a diagram, the Hertzspring-Russell diagram which relates the absolute luminosity against the spectra of different stars. In the diagram most stars fall into a narrow band, and are called the Main sequence stars. At the top are bluish white stars whilst at the bottom are duller reddish stars The colour indicates the surface temperature. The bluer stars are hotter. The hotter the temperature the more massive is the star.

The range is up to about 100 solar masses. Such large stars quickly burn out, explode as a supernova (Type 2) and go on to form black holes.

Surprisingly the approximate mass of the central black hole in a spiral galaxy can also be determined in principle by the same method and relies of an estimate of the mass of a star orbiting the black hole. If they have the same colour as our sun they are assumed to have the same solar mass. The orbital velocity can be determined from the red shift of that star and an approximate estimate made for the distance from the centre of the galaxy. Although this is a gross simplification the results indicate that the mass of any central black hole in a spiral galaxy can run into millions of solar masses.

The mass of a galaxy can be inferred by estimating the number of stars by sampling various areas. This can only be done for nearby galaxies and the results are vary crude but indicate that our own galaxy could have a billion or more stars, the commonest being those like our sun. Our galaxy, and our neighbour, the Andromeda galaxy are approximately the same size and seem to be a common size for moderately large spiral galaxies. But galaxies vary in size.

The composition of a star can be determined by its spectrum. When heated each element has it own particular pattern of a set of frequencies of the light radiation. Analysing the frequency bands enables different atomic elements to be identified. It was from this when applied to the sun that helium was first identified, long before any pockets of helium were found on earth. This method does not identify elements deep in the heart of any star although convection cells are likely to bring up to the surface most elements that are in the heart of the star.

There is no method of identifying quantitatively the amount of each element in any star although a rough approximation can be obtained as to which are the commonest. By far the commonest are hydrogen and then helium. When the surface concentration of hydrogen falls below 50% this is a sign of trouble. That is this star is about to die, either as a type 2 supernova or else inflate itself into a red giant and regenerate itself as a helium burning star.

Chapter 3

The Hubble Wars and the fifth force

Hubble's constant, its derivation, the theoretical basis for the Hubble constant, observational proof, a fifth force, the fuel sources of gravity and expansion, quantum considerations, life of mass. Two fundamental constants, the defining proof.

In the 1930's Edwin Hubble, the doyen of American astronomy, made two momentous discoveries. These were that the universe is expanding and that there was a constant relationship between the velocity of the receding galaxy and its distance, the greater the distance the higher the velocity. This became known as the Hubble constant. He was only able to determine its value from comparatively nearby galaxies as the significance of the Cepheid variables had not been fully realised. These are stars, which fluctuate in brightness in a regular pattern, the period of the cycle is directly related to the maximum brightness of the star. Cepheids are moderately common in any galaxy. Within our galaxy it proved possible to obtain the absolute brightness as it related to the periodicity. Identification of the length of the period of a Cepheid in a distant galaxy enabled its observed brightness to be matched with its absolute brightness to determine, using the inverse square law, how far away it is. This technique was not available to Hubble. At the time the unit of distance that he used was the parsec, or parallax arc second, which equals 3.26 light-years.

His formulation was that the velocity of a galaxy in kilometres per second divided by the distance in Mega parsecs was a constant which he put at around 500 km/sec/Mega parsec. Some sympathetic understanding is necessary since he did not have an accurate method of identifying distances, and a lower figure implied a size of the universe which was way outside all contemporaneous estimated values. At that time in American Science there was a considerable reluctance to accept that nature provides huge figures. This was also reflected in earlier considerations of the energy of the atom. Such big figures reduced man's place in the universe to that of extreme insignificance and were therefore socially unacceptable. Happily the acceptance of big figures is now normal.

Since Hubble's discovery there has been considerable argument about the precise value of the constant. With the improvement of measuring distances the 500 figure was quickly reduced to around fifty km/sec/Mega parsecs. But there have been great disputes about this. This argument has led to what some have called the Hubble Wars. Some French astronomers put the value at around 100km/sec/Mega parsec. American astronomers have stuck firmly to the approximately 50km/sec/Mega parsec value. More recently NASA has issued what appears to be an *Ex cathedra* statement, that it is 70 km/sec/Mega parsec. Gribbin, in a monumental analysis of nearly 1000 fairly nearby galaxies, found the mean value was 52 +/-6 km/sec/Mega parsec. The driving force behind all this squabbling is that one can use the Hubble constant to determine the age of the universe by simply back tracking until all the galaxies are at one place. Calculating how long that would take would give the age of the universe.

One of the surprising things about Gribben's value, of 52 +/-6 km/sec/Mega parsec from a set of nearly one thousand observations, is the size of the standard deviation of the mean, +/-6. For a physical measurement based on such a large sample size this is a very large value. The calculation of Hubble constant for any individual galaxy is based on two data points, the velocity, which can be measured with great precision using the red shift, and the measurement of distance. This is normally taken to have an error for any individual reading of +/- 10%. (The standard deviation of the estimate) The standard deviation of the mean is this value divided by the square root of the number of measurements in the sample. 10% of fifty-two divided by the square root of 1000 yields a value of 0.15. That is the scatter in Gribben's analysis is more than forty times what one would expect on a normal distribution, which should have just the scatter due to observer error. Gribben's large scatter carries a great implication. Some other factor is at work disturbing the mean value of the constant that the statistics were trying to establish. Further this effect is not random. In particular it suggests that there is occurring a systematic variation in the value of this mean over the range of the distance measurements in this sample of a thousand galaxies. That is to say the value of the Hubble constant was changing across the distance range of his sample.

But the unit Hubble proposed dimensionally does not, in physical terms, make sense. It is distance per second divided by distance. This results is a rate of change of a dimensionless number, X per second (in dimensional analysis the two distance dimensions cancel out). The numerator does not refer to anything. It is like saying pi per second, or describing a frequency.

Despite the Hubble constant not making any sense in physical terms, that is in dimensional terms, it showed a relationship between the velocity of a galaxy and its distance. Since its discovery, the constancy of the Hubble constant had been assumed to be applicable across the entire universe. The constant, subject to the correction resulting from more accurate means of determining distances, has been checked many times since the initial observations were made. But all these confirmations were based on nearby or relatively nearby galaxies. Various techniques have been used e.g., by using the differential velocity of the stars that form the outermost parts of elliptical galaxies. Oddly these same measurements have since been used to claim the existence of so called dark matter.

What does not seem to have been appreciated is what the Hubble constant signifies in terms of normal physics. It is a matter of regret that as the denominator of his constant, Hubble used the rather strange unit of distance he was promoting, the Parsec. This obscured the true significance of his observations. Had he used light years as the unit of distance it would have quickly become apparent that he was describing an acceleration constant. Clearly also if the velocity increases with distance then as the galaxies recede they must be being accelerated. The problem obviously was that based on his value for the constant, the age of the universe would have been less than the age of the dinosaurs. Such a controversy would have obscured the true significance of his findings, that the universe was expanding. Up to that time it was assumed that the universe was static. Indeed Einstein had to build in a constant to correct for any expansion that his equations were predicting, a matter he subsequently regarded as his greatest mistake. Had Hubble re-examined his thinking and conclusions, once he had available improved means of determining distances, he would have found that he had identified a fifth force in nature, comparable to but greater than gravity. His standing in the pantheon of great scientists would then have been comparable to Newton who had identified the fourth force, gravity.

If the Hubble constant is 48 km/sec/Mega parsec this then becomes 48/3.26 km per sec, or 14.7 km/sec, per million light years. If however one substitutes the time taken for light to travel those light years it becomes velocity per unit time which is acceleration. The value of the time adjusted constant then becomes 4.66×10^{-10} metres/sec/sec. If the Hubble constant has a value of 70-km/sec/Mega parsec replacing the Mega parsecs with time the acceleration constant becomes 6.80×10^{-10} metres/sec/sec. These values can be contrasted with the Gravitational constant G which has a value of 6.67×10^{-11}. That is H_T, the Hubble time constant, or acceleration constant, is approximately seven or ten times more powerful than G the gravitational constant. It is small wonder that the universe continues to expand. What Hubble had inadvertently discovered and had failed to realise, was no less than a fifth force in nature. One that should be added to the strong force and the weak force, that operate within the nucleus of the atom, the electromagnetic force which governs electromagnetic radiation such as heat and light, and gravity.

One unfortunate by-product of using parsecs is that it directed astronomical attention away from considerations of the mechanism. It also prevented a more detailed evaluation of any theoretical basis that underlies the constant, a basis that could be used to construct an over arching physical theory of the expansion of the universe. Hitherto it has not been possible to account for the expansion, and the fact that very distant galaxies are receding faster than nearby ones. Tinkering with the energy of the Big Bang to explain this is reminiscent of the use of epicycles to describe the motion of the planets that Copernicus with his heliocentric views swept away.

The source of this expansion force must come from the molecules that make up the galaxies. If there were any external force the galaxies would not have been able to maintain their structure as spiral or elliptical galaxies. Galaxies are a mixture of matter from the densest and most massive, such as black holes, to the thinnest of nebulous clouds of dust, all loosely held together by gravity. It follows that each component of the galaxy emits a force that is proportionate to the mass of the component, so that the whole galaxy is accelerated in unison. In accelerating the various masses work is done, and energy is being expended. There must be a continuous source of that energy to replace what has been used. The amount of energy required for this replacement is enormous.

Unlike gravity, which is a force, uniting various masses, the acceleration force from one mass appears to be independent of any acceleration force coming from another mass. Its function seems to be to accelerate the mass in the direction it is travelling. (For a detailed possible explanation of the mechanism behind this see appendix 4 at the end of this chapter). As a consequence galaxies are accelerated away to distant space, creating the expansion of the observable universe. Within galaxies it adds a small boost to stars that are in orbit around the central core of the galaxy. For stars near the core of a galaxy this boost is small relative to the orbital velocity which results from the gravitational force exerted by the core. But on the periphery of a galaxy where the central gravity effect is weakest this weak acceleration effect produced by this force becomes more noticeable. Stars at the periphery of the galaxy will move faster than the speeds predicted by the standard equations of orbital mechanics. Without this extra velocity the spokes of the spiral galaxies would be much more angled than they are.

This would seem to be a key function of this acceleration force. Planets, such as earth orbit their parent star in an elliptical fashion. At perihelion when the earth is nearest to the sun it is travelling faster than when it is furthest away at aphelion. The acceleration is provided by the sun's gravitational pull. As the earth swings past the sun travelling towards aphelion the same force decelerates it. It thus behaves as a machine going through a cycle of acceleration and deceleration. The earth has been doing this for billions of years, as have all the other planets in the solar system, and by extension all those planets that orbit other stars. The constant acceleration and deceleration means work is being done and at apparent 100% mechanical efficiency for all this time. It is a perpetual motion machine. But this is impossible. The laws of thermo-dynamics forbid this. Meanwhile, as will be shown in later in this book, the mass of the planets is slowly lessening, as also is the mass of the sun, because of a time effect. The ensuing gravitational consequences cancel out. But the sun is also losing large amounts of mass as energy heating up the rest of the solar system. Its gravitational strength is also weakening, albeit extremely slowly.

Because of the size of the earth's orbit there will also be relativistic consequences as the reduced gravitational force from the sun would take time to reach aphelion. That is there must be some mechanism or other by which the orbit is being constantly maintained. With the loss of energy from a less than perfect mechanical efficiency the system should slow down. But if the velocity slows down the planet's orbit must change, and the earth should spiral down into the sun. Yet paleontological records show that the earth's temperature has been remarkably constant for at least half a billion years (apart from blips from assorted ice ages). The geological record suggests that a stable temperature has been there for much longer. It is suggested that the small acceleration force arising from earth's mass compensates for this decay in orbital velocity. The same applies to all the planets in the solar system, and indeed all-orbiting bodies that are at a distance from their axis of orbital revolution.

Black holes though raise a conundrum. Their gravitational strength is so great that forces such as the electromagnetic force cannot escape. Yet clearly black holes are accelerated along with every other part of the galaxy. Furthermore because of the immense gravitational field that is produced immediately around the black hole time is slowed. In any zone of time slowing force is attenuated when viewed from outside. Time slowing causes mass to shed as energy. Some of that energy will be gravitational, but some of the energy will be in the form of acceleration. This would compensate for any attenuation in effect caused by time slowing. The increased gravitational energy will further slow time and so the process must be continuous. Thus a black hole is able to keep pace with the rest of the galaxy as that galaxy is being accelerated through space. This is, in fact, an over simplification of the problem of accelerating black holes—as is discussed in a later chapter of this book.

As will be seen later in this book it is proposed that time is slowing down. The period of the second is now twice as big as it was when the visible universe was half its present size, that is had half its present radius. The basic constants of nature, such as c the velocity of light, G the gravitational constant and including the Hubble acceleration constant adhere to the prevailing pace of time. In other words they are constant in all time frames. It follows that when we look at very distant galaxies we should see their acceleration in their time frames. That is, if uncorrected, the Hubble "constant" should increase with distance. The ratio of their age to our age (in earth time) is the ratio of their second to our second and so can be used to predict the Hubble value that is to be expected with increasing distance.

There is a caveat. The acceleration constant H_T is defining a rate of change of movement. It is in direct conflict with gravity which also defines a rate of change of movement. But the gravitational acceleration depends upon two masses and the distance separating them. It follows that when the gravitational acceleration is substantial, or to put it another way when the local density of gravitons is very high, this will interfere with the actual acceleration the expansion force is trying to produce. It is only when the distances are very substantial, that this gravitational effect will be negligible. It is tentatively suggested that the slight downward trend of the constant shown in Figure 4,1 might be due to this kind of interference but more evidence on the distances and velocities of many more very far distant supernovas will be needed to clarify this point. The key relevance of this is its application to the early expansion of the universe. This is dealt with in Chapter 10

The theoretical basis for the Hubble constant and the consequences

The Hubble constant is the ratio between velocity and distance. Although it is supposed to be constant to the edge of the universe clearly this cannot be so if there is a maximum velocity, c the speed of light. If the universe continues to expand when its edges are expanding at the speed of light then the ratio will become progressively smaller.

It is worth considering the basic geometry of an expanding sphere and the consequences of this. This is shown in Figure 3.1 and the analysis is given in detail in Appendix 1 appended at the end of this chapter. What emerges is the simple equation

(Equation 3.01) $$H_0 = \frac{v}{D} = \frac{652}{Age-D_T}$$

where v is the velocity in km/sec, D is the Distance in Mega parsecs, Age is the Age of the universe in billions of years in earth time and D_T is the time taken for light travel the Distance D and also is in units of billions of years.

What is of interest is that equation allows predictions of distances to be made if one knows the velocities and assuming the age to be around 13.5-13.8 billion years. These can be compared with numbers calculated from observational data where the distance is known with reasonable accuracy.

The significant thing about this is that the Hubble constant is not constant but must vary with the time taken (in billions of years) for light to travel from the galaxy in question. If for any group of galaxies this time is small relative to the age of the universe it will appear that the Hubble ratio is constant. It was this that has led to the extrapolation error in assuming that the Hubble constant was constant throughout all time and space when only nearby galaxies were looked at.

Proof that the Hubble constant is not constant

From what has been described it should not be difficult to establish whether or not the Hubble constant is truly constant.. The records of 80 galaxies, and supernovas were used. The data and their sources are listed in Table 3.1. The data were arranged in order of velocity. Data on their distances was either calculated from the magnitudes of the supernovas, or, for the nearby galaxies, from older more established means. The distances covered up to almost 7 billion light years or almost 3000 Mega parsecs.

The supernova data suffers from one difficulty. The absolute magnitude or brightness of Type 1a supernova varies with time and it also varies with the wavelength of the emitting radiation. The spectrum of a supernova shows a mixture of wavelengths each of which changes as the supernova expands and cools. Both require compensation and a certain amount of extrapolation to identify peak brightness. From there on it is simply a matter of the inverse square law comparing the derived brightness against a hypothetical supernova of the same type at a standard distance.

There is no difficulty with the latter. For the very far supernovas whose magnitudes are very large (that is they are very dim) a small error in the assessment of magnitude has a disproportionately large effect on the estimation of distance. This is because magnitudes are on a negative logarithmic scale with each unit change of magnitude representing a two and half fold decrease in brightness of what went before. The magnitude and distance scales shown in Figure 3.3 illustrate this point.

From the above, galaxies of similar velocities should be at similar distances from us. Where there is a discrepancy the implication is that there has been some interference with the light from that particular supernova. Thus if the supernova occurred deep in the heart of a galaxy, there could well be a cloud of dust, its equivalent of the Horse's Head nebula, partially obscuring the light. Equally if the host galaxy of the supernova has a cloudy halo this could also lead to an over estimate of the magnitude of the supernova. This effect would be more pronounced from any dusty halo surrounding our Galaxy as the light from the supernova would already be extremely dim, that is any interference or adsorption of the light would have a disproportionate effect on the estimate of distance.

The sources of error when determining the distances of these dim and distant supernovas give rise to a skewed distribution of the scatter. Small patches of dust, whether within the host galaxy or in our Galaxy, could obscure some of the light and so lead to an over estimation of the magnitude, and hence an over estimate of the distance. In contrast magnification of the light, as by a gravitational lens could lead to an underestimate of magnitude and place the supernova nearer than it should be. But gravitational lenses are much rarer than patches of dust.

One other source of error that applies to the most distant and therefore the fastest supernovas, is due to their high velocities. There are relativity effects on the magnitudes. The magnitudes or brightness of standard candles means that the rate of release of photons of appropriate amplitude (i.e. the photons summate or add up) is also standard. But rate means it is time dependent. If time slows for any reason then judged from a distance in normal time (or in the jargon of relativity then relative to us) the candle will appear less bright. Its magnitude will be greater. But it is straightforward to compensate for this. (See appendix 2).

Figure 3.2 shows the results for some 40 galaxies all within a billion light years distance, comparing distance with velocity. It was from data such as this that Hubble and others since have assumed that the Hubble ratio is constant and the constancy can be extrapolated to the edges of the universe.

Figure 3.3 shows that this is not the case. Distance data, listed in Table 3.1 derived from 80 galaxies and distant supernovas, up to ~8 billion light years, were compared with the predicted value. The Hubble ratio increased with distance and the predicted values for distance matched very closely (within the error of measurement of such distant objects) the values derived from observational data.

The conclusion is that the Hubble constant is not constant but varies with distance that in the past astronomers have extrapo-lated the value to cover the universe but it turns out their extrapo-lations were unjustified.

All of this sounds very esoteric, like arguing how many angels could fit on the point of a pin. But it carries a profoundly important message, one that strikes at the very heart of cosmology. This is the behaviour of time. And even more importantly the relationship between time and mass and the consequences of this behaviour. This will be dealt with in the subsequent chapters.

The fifth force

As was discussed earlier in the chapter the Hubble ratio could be described as an acceleration constant that is being disturbed by time (The case for this is discussed in the next chapter). Changing the denominator of the Hubble ratio from distance to time results in an acceleration factor of ~5 x 10^{-10} m/s/s. That is a force, which has this acceleration factor and so generates this expansion, is driving the expansion of the universe. It clearly works against gravity which otherwise should cause the universe to contract. Its strength accounts for the expansion of the universe even when the current observable universe was very small and essentially the universe was a little over the size of one very very large black hole.

It clearly arises from the mass that is being accelerated. Galaxies are a mixture of masses of different densities which vary from being as thin as a nebula to as dense as a black holes, albeit the masses are loosely held together by gravity. Any powerful external force (and to move galaxies it must be powerful) would disrupt the geometry of the galaxies. Yet galaxies move with its components in lockstep with each other. This means that this fifth force, which is arising from the mass being accelerated and which force is in exact proportion to each of the component different masses within the galaxy, is the explanation of the so-called dark energy hypothesis.

One of the current cosmological theories is that there exists something called dark energy. It is claimed that this energy is responsible for the expansion of the universe, and labelled it dark because they were unable to define its nature or origins. Elaborations on this theme have proposed that there are two kinds of dark energy, positive where it is repulsive to gravity, causing the expansion of the universe, and negative where it imitates gravity. Positive dark energy it has been suggested is gradually weakening. It arose or is a property of a fixed volume of vacuum. As that vacuum expands (or space expands) so it is weakened. This was initially called the cosmological constant. Variations of this gave flavours to dark energy, the quietest being vanilla, the most exotic being quintessence while the most vigorous posits that dark energy is ever increasing and threatens to rip the whole universe apart. This has been called the phantom menace. What negative energy is in physical terms is somewhat obscure.

This hypothesis would have it that the energy for expansion arises from some form of degradation of space, much as quantum theory postulates that pairs of electrons and positrons random emerge spontaneously and very briefly out of space. They annihilate each other with the emission of energy in the form of a photon. This contradicts a fundamental tenet of physics that energy cannot be created or destroyed. Alternatively if space is a form of energy, that was created at the same time as the Big Bang, these virtual particles imply that space is being reduced.

A force emanating from the mass itself, much as gravity emerges out of mass, poses a much simpler solution. Clearly this force, in moving galaxies against the force of gravity consumes lots of energy and the source of this and the mechanism for its release is discussed in a subsequent chapter.

But there are some unusual confirmations of this fifth force. The Swiss astronomer Zwicky more than fifty years ago observed that in-groups of galaxies some of the outer galaxies were moving faster around the group than could be accounted for by the gravitational force exerted by the other galaxies. He postulated that there must be some large unobservable mass whose gravitational force was responsible. He coined the term dark matter.

Since then it has been noticed that stars on the very edge of galaxies are moving some eight times faster than predicted from their position in the galaxy and their distances from the centre of the galaxy. The equations relating gravity and mass to orbital velocity predict there is mass which is more than eight times the mass of the galaxy as determined by the orbital velocities of the inner stars in that galaxy. The hidden assumption is that only gravity can cause the orbital velocities. This has led to the re-emergence of the concept of dark matter exerting a gravitational force. Why its effects should be confined to the perimeter of a galaxy is not explained. If there is so much mass concentrated on the perimeter of the galaxy it should have powerful effects on the next innermost layer of stars and cause lots of black holes to form. But this is not what is seen.

M. Milgrom has calculated the velocities of these faster moving outer stars and suggested that they are being subject to a force of $\sim 10^{-10}$ m/s/s and this force is an acceleration force. For want of a better explanation he attributes this to a deviation of the normal role of gravity changing with distance, that is Newton's theory of gravity has a slight error. The fifth force, the acceleration force identified from the Hubble constant data would eliminate the need for any dark matter and leave the geometry of the galaxies as well as Newton's theory of gravity unchanged.

That does mean the main body of the galaxy and its central black hole exert the gravitational force that determines the orbits of most of the stars. It is helped by the acceleration force, but for these fast innermost stars the additional velocity imparted by the little acceleration force is hardly noticeable. It is when the gravitational force itself is very weak, as at the edge of a galaxy, the acceleration force effects become noticeable. This is the H_T acceleration force, which does not become obvious until the stars are a very long distance from the central black hole. This is because the pure gravitational force is subject to the inverse square law and is very powerful close to massive bodies like black holes. The acceleration force does not interact with the other stars it merely acts on the body producing it to accelerate it (and hence causes black holes to accelerate *parri passu* with the rest of the galaxy). What perhaps is surprising is that Allan Sandage, Hubble's successor, used these same stars to calculate the Hubble "constant" and came up with a result that closely matched established values. He found that value was 43 +/- 11 Km/s/M.parsec. Substituting time for megaparsecs gives 4.2+/-1 x 10^{-10} m/s/s as acceleration.. His figures confirm Milgrom's calculations. But if the accelerations of the galaxies and the accelerations of these outlying stars were due to the same force, this result would be expected. But it serves to confirm the demise of the dark matter hypothesis.

Then there is another mystery. In 1972 NASA launched a space probe Pioneer 10. A year later it launched Pioneer 11. Now more than thirty years later the two spacecraft are 400,000 kilometres further away than expected. They have been subject to an unknown acceleration force that is described as less than a nanometre per second per second i.e., $< 10^{-9}$ m/s/s. Half a nanometre per second per second is 5×10^{-10} m/s/s. NASA cannot account for what to them is an anomalous acceleration. Yet it is exactly what is to be expected if the probes were subject to the Hubble expansion energy. Attention is drawn to two numbers 4.86×10^{-10} m/s/s, which is the H_T value if the Hubble constant is 50 m/s/M.Parsec. and 6.8 x 10-10 m/s/s if the Hubble constant is 70 m/s/M.Parsec. The two Pioneer probes provide the perfect experimental framework for determining the exact value of the Hubble constant.

Thus there are three sets of bodies of vastly different masses, the galaxies, the outer stars of the galaxies and the small Pioneer probes are all showing approximately the same acceleration. With the galaxies there is a small variation with distance, which is accounted for in the next chapter. But clearly this uniformity of acceleration cannot be a coincidence. The unifying concept is a fifth force that is additional to the four fundamental or primary forces of nature, the electromagnetic force, the intra nuclear strong and weak forces and gravity. But such fundamental forces have their specific constants, which are constant relative to the time frame. For the electromagnetic force this is c the velocity of light, for the Gravitational force it is G the gravitational constant. And so on. The variation in the acceleration force with distance strongly suggests that in the past the time frame was different. This aspect is explored in detail in the next chapter.

But it does not stop there. In Germany Burhhard Heim, a theoretical physicist in the 1950s started working on a unified field theory linking relativity theory and quantum theory. He was eventually successful in that he was able to predict the masses of a large number of subatomic particles. When these theoretically derived values were compared with those obtained by measurement the results were virtually identical. But in his theory there emerged two fundamental forces of nature that had not been described or indeed known before. One of them has been characterised as being antigravity, that is the force ignores gravity, it arises from the mass being accelerated and is responsible for the acceleration of galaxies and expansion of the universe. This theoretically derived force matches extremely closely the observational derived force, H_T. The sole point of discrepancy is that the observationally derived H_T varies with distance. If it was a fundamental force of nature it should be constant. The case for a fifth force would appear to be irresistible

There is an odd historical coincidence. The planet Pluto and its orbit were determined on purely theoretical grounds long before the planet was identified and orbit confirmed. Similarly the occurrence of a fifth fundamental accelerating force was determined on theoretical grounds long before it was identified for what it is. Is this a case of history repeating itself?

The fuel sources of Hubble expansion and gravity

A force that causes movement does work and so uses energy. If that force is to be maintained then the energy must be replaced. Of the five forces of nature (including the acceleration force) the strong force within the nucleus of the atom causes negligible movement and so generates almost no energy deficit. The weak force uses some energy in expelling helium nuclei from radioactive atoms. It derives its energy from the radioactive decay from the larger atomic nuclei. The electromagnetic force dissipates its force; for light the energy is mainly used in such things as the energy of photosynthesis, or radiant heat, or simply propagating powerful electromagnetic waves, from radio waves to Gamma rays. The electromagnetic force transmits a lot of energy and that has to be replaced. The energy required for this is derived from changing the orbits of the atom's circulating electrons as they sink into a lower energy level on releasing the energy as photons. Other electromagnetic energy is derived from the loss of mass on atomic fusion, most commonly hydrogen to helium fusion that occurs in stars such as our sun.

But the energy used by electromagnetic radiation pales into insignificance compared with the energy expended by gravity. Gravity stops bodies from disintegrating. It holds the oceans, the sand and the soil on to the earth. Newton's laws of motion would predict that these loose particles would continue in a straight line unless pinned down or acted upon by an external force. Gravity changes the direction of movement of the earth so that it orbits around the sun. This happens with all the planets in all the various stellar systems that exist not only in our galaxy but also in all galaxies. It is responsible for the orbit of the sun together with all the other stars as they orbit around the central core of our Galaxy. And the same applies to other galaxies. It pulls galaxies together and on occasion causes them to collide. The energy used to do all this is prodigious. In energy terms just in our own Galaxy it is the equivalent of many millions if not billions of tons of mass being reduced to energy every second. And this has been going on for billions of years.

But mightier yet is the acceleration force, which is responsible for the expansion of the entire universe. Every galaxy is being accelerated away from the site of the Big Bang and that requires energy that is an order of magnitude greater than used by gravity. It is this that is called, by some, dark energy.

Such large demands for energy can only be met by mass reverting back to energy. Furthermore this reversion must occur within all masses, whether they are very dense or very tenuous, and must be as a fixed proportion to the quantity of the mass. Further there must be some more or less universal controlling mechanism whereby the energy produced relative to the size of component of the galaxy is the same for nearby galaxies and the most distant. The energy released does work in moving the mass from which it originates. This movement is by accelerating the mass either towards another mass if it can, as gravity does, or simply accelerating the mass in the direction of movement of that mass. The transformation of some of the mass as energy is a very quiet and slow process distinctive from the more readily apparent and very violent transformations that occurs in the inner core of stars (and atomic weapons). There is only one mechanism available whereby such large quantities of mass can be transformed back to energy, yet per unit quantity of mass only a minuscule amount of mass is shed per second. How this occurs is discussed in a later chapter

Balancing the books or the life of mass

The equations generated when auditing of the books of any system frequently unearths some surprising conclusions. This is very true when one audits the energy accounts of the expansion of the universe. It is axiomatic that the amount of energy within the universe is finite but very large. Energy cannot be created or destroyed although it can be changed in form, as Einstein's famous equation shows. There is a small caveat, quantum theory proposes that popping out of space are pairs of virtual electrons and positrons. They mutually annihilate each other unless near some very massive object which can drag the two virtual particles apart. The mutual annihilation results in the production of photons. But photons are packets of energy. The question then arises where does this energy come from. There is as yet no estimate of the mean rate of the production of virtual particles per unit volume per unit time. There is no proof that this actually occurs. It is therefore assumed to be very small and for the purpose of this book keeping exercise can be ignored.

It takes energy to accelerate a mass. The energy for accelerating a galaxy comes from the mass of the galaxy, with each component part of that galaxy contributing according to the mass of that part. In effect the mass is burning itself up in order to provide that acceleration. One can calculate the distance that will have to traverse when all that mass has been used up (See technical notes at the end of this chapter). If the acceleration (the Hubble expansion constant) is 4.86×10^{-10} m/s/s this extinguishing distance is 19.5 billion light years. If one takes NASA's value for the Hubble constant this distance becomes 14 billion light years

It is also possible to calculate how long it will take, whilst producing the energy required for acceleration, before mass extinguishes itself. Incidentally this defines the life of a proton. Using the 4.86×10^{-10} m/s/s the figure is 27.6 billion years. Using NASA's value for the Hubble constant this figure is 20 billion years.

There is a remarkable coincidence from the world of atomic physics. For various reasons totally unrelated to the above discussion some atomic physicists have concluded that the life of the proton is between 25 and 30 billion years. Could it be that the life of the proton is determined by the slow leakage of energy, energy that is used to accelerate that proton, and indeed all mass in an expanding universe? In making their measurements have the atomic physicists stumbled on experimental proof of the acceleration constant? It makes a fascinating question.

To sum up. The Hubble constant cannot be constant over the whole universe. Replacing the denominator of the constant by time shows it is an expansion force that is a hitherto unrecognised fifth fundamental force of Nature. It accounts for the anomalous velocities which have **been** attributed to dark matter and dark energy. The fuel for the acceleration force arises from the mass being accelerated. This imposes a limit on how far the mass can travel before it uses up its substance. And also enables the life of the mass to be determined.

The probability is that energy, under conditions of high temperature and pressure will form new mass, that is form new protons and neutrons. This replicates the situation at the time of the Big bang when mass first formed. Such conditions would exist in the midst of a type 2 supernova. Such a cycle would prolong the life of a galaxy

Appendices and Technical Notes

Appendix 1 The theoretical basis of the Hubble constant, a geometrical analysis.

If the edge of an expanding sphere is moving at a more or less constant velocity (over a period of one or two seconds), of i unit of distance per second then the mean change of that distance for those seconds is i/2 per second. If the sphere is radius R the volume V of the sphere before that second was

(Equation 3.02) $\qquad V = \dfrac{4\pi R^3}{3}$

The volume for the next second averaged

(Equation 3.03) $\qquad V = \dfrac{4\pi(R+(i/2))^3}{3}$

The volume for the previous second averaged

(Equation 3.04) $\qquad V = \dfrac{4\pi(R-(i/2))^3}{3}$

But from Figure 3.1 standard geometry shows that the ratios agree

(Equation 3.05) $\qquad \dfrac{i}{2R} = \dfrac{v}{2D}$

where $v/2$ is the mean velocity for a point, e.g. a galaxy, or when applied to the expansion of the universe, is the velocity of a galaxy at a particular time when the point or galaxy is at D distance.

(Equation 3.06) $\qquad \dfrac{i}{2} = \dfrac{Rv}{2D}$

When this is substituted in equation 3.03 and the equation is expanded it becomes

$$V = \frac{4\pi}{3} (R^3 - \frac{3R^2Rv}{2D} + \frac{3RR^23v}{4D^2} \text{ and so on.})$$

But these latter parts are so small particularly when D the distance is very large that they can be ignored That is for t seconds

(Equation 3.07) $\qquad V = \frac{4\pi}{3} (R^3 - \frac{t3R^3v}{2D})$

It follows that when the Volume was essentially zero (the time of the Big Bang)

(Equation 3.08) $\qquad 0 = \frac{4\pi}{3} R^3(1 - \frac{T3v}{2D})$

where T is a large number of seconds, v is the velocity of a galaxy expressed as a fraction of the velocity of light, and D is a large number of light seconds (this makes the arithmetic extremely simple). Swapping seconds for billions of years or light years results in

(Equation 3.09) $\qquad T = \frac{2D}{3v}$

But T is the time taken from the Big Bang to the time the galaxy shone its light at us. It then takes more time for that light to reach us. This is D_T is the transit time. T and D_T together add up to the age of the universe.

Put another way $\qquad T = \text{Age} - D_T$

Inverting equation 3.09 gives

(Equation 3.10) $H_o = \dfrac{v}{D} = \dfrac{2}{3(Age-D_T)}$

When v is converted from fractions of the velocity of light to kilometres per second and D is converted from billions of years to Mega parsecs the value of the Hubble ratio is then

(Equation 3.11) $H_o = \dfrac{v}{D} = \dfrac{652}{(Age-D_T)}$

which is equation 3.1

If the Age of the universe is 13.7 billion years the Hubble value is 47.6 km/sec/M.parsec. This theoretical value may be compared with the mean time corrected value of 49.2 km/sec/M.parsec derived from 80 galaxies listed in Table3.1

Appendix 2. Calculation of distances to very remote supernovas.

There were a wide variety of sources of the data used in this work. This covered data on galaxies and supernovas that were at distances up to almost three thousand Mega parsecs, that is more than seven billion light years. The details on 80 supernovas and galaxies are given in Table 3,1 The sources of the information are detailed in the notes attached to the table. The majority were from NASA's Supernova project. Otherwise where the distance to the galaxy or supernova was given this was taken. Where the distance had to be calculated from the magnitude this was derived using a slight variation on the formula described by Kayer R. et al. (Kayer R. Heibig P. and Schramm T. (1997) A general and practical method for calculating cosmological distances. In Astronomy and Astrophysics 318., 680-686.) The formula relies on the inverse square law and takes into account that magnitude is a logarithmic scale with the base 2.51. The formula is

(Equation 3,12)   Log Distance = (M +m+k-r-25)/5

Log Distance is in the standard logarithm, base ten, units and when the antilogarithm is taken the answer is in Mega parsecs.

M   is the standard magnitude of a Type 1a Supernova that is at a distance of 10 parsecs. It has a value of –19.7.

m   is the peak magnitude of the individual supernova listed in Table 3.1

k is a standardised correction factor, that is listed also in the table. Details about the k factor can be found in a paper by Coleman ((Coleman G.D., Wu C. - C., and Weedman D.W. (1980) ApJS 8, 854). A more detailed analysis of the k factor is given in a chapter by Sandage A. (1995) entitled Practical Cosmology, Inventing the past. in the book The Deep Universe, (Eds. Binggeli B. and Buser R. Published by Springer, Berlin).

The 25 and 5 numbers are adjustments necessary when the answer is to be in Mega parsecs.

r is a new term to denote the correction to the magnitude necessary because of the relativity induced time dilatation arising from the high velocity of the supernova. The rationale for this is that the brightness of a type 1 supernova follows a very set pattern over a period of time. The light is the result of specific nuclear events and the rate of the emission of photons is a consequence of this. Hence if time is slowed for whatever reason the rate of emission of photons will, to a distant observer, apparently be less and this will be reflected in the brightness or magnitude. Thus if time is slowed to half normal the brightness will be halved. The relativistic expansion of time can be calculated from the velocity using the standard relativity equation relating velocity and time. In simple integers this is a multiplier with a value of $(1-v^2)^{0.5}$ where v is the velocity of the galaxy or supernova expressed as a fraction of the velocity of light. After adjusting for the base of the logarithmic magnitude scale is the relativity correction in magnitude units

(Equation 3.13) $r = 2.51 \times 0.5\log (1/(1- v^2))$

Table 3,1 shows the data including the resulting r values from 80 galaxies or supernovas. It is displayed in increasing order of size of the velocity. This enabled aberrant values to be identified, in that those objects of the same or similar velocity should be at the same or similar distance. Causes of aberrant values of stellar magnitude could be the interception of some of the supernova's light by dust. That dust could be from the host galaxy or in our galaxy. If the stellar magnitude is unduly low this suggests that the light was being amplified, as by some form a gravitational lens. It was also obvious that one supernova, Number 49, SN1994G had an unusually large k value correction. From the original data it also had a very substantial uncertainty as to the value of its peak magnitude and it resulted in the highest value, by some margin, for the Hubble constant and time corrected Hubble constant. Nevertheless this value was accepted for inclusion in the statistics.

The raw Hubble constant value was calculated, dividing the velocity in kilometres per second by the Mega parsec value. It was found, as anticipated, if time is expanding that the value of the Hubble constant, the slope of the line, did indeed rise with increasing distance. Figures 3,2 and 3,3. That is that the value of the Hubble constant is only constant for relatively nearby galaxies and supernovas, when it could well be slightly above the true mean. It is small wonder that the original observers thought that the Hubble value was constant, and by extrapolation constant throughout the universe. It was also noted that for nearby galaxies the Hubble constant was very close to 50km/second/ Mega parsec.

These observations could also well account for NASA's declaration that the Hubble constant has a value of 70km/M parsec in that they were looking at rather more distant galaxies. But equally beyond those the Hubble value rises even further. Figure 3.3 also shows an increasing apparent scatter as the distances increase. The figure also identifies the difficulty in obtaining the exact magnitude of the distant supernova in question. At magnitudes of 22 and beyond a 1% error in magnitude estimation (i.e., 0.2 magnitude error) could lead to an error in distance calculation of several hundred Mega parsecs, or a billion light years. Nevertheless the errors that did occur were clearly random and the predicted velocity for any distance (the heavy curved line in Figure 3.3) passed midway through the scatter.

Also shown in Figure 3.2 is the expected velocity with changing distances at various putative Hubble constants, 50,60,70 km/sec/ M.parsec. It is apparent that initially there was little to choose between the three values and so it is not surprising that there is dispute as to its exact value, although underlying this argument is the assumption that the value is constant throughout the universe.

Appendix 3. The extinguishing distance and extinguishing time

The general relationship between mass and the energy required to accelerate that mass is

(Equation 3.14)   Energy = mass times acceleration times distance

Since distance is acceleration multiplied by time divided by 2 or

(Equation 3.15)   Energy = mass x acceleration^2 x time/2

These are the normal equations relating energy to acceleration. The difference with Einstein's famous e $=mc^2$ is that at very high velocities according to his special relativity effect the time is expanded so that the divider 2 disappears

But from that Einstein equation energy (in Joules) when divided by 9×10^{16} is mass (in kg) as. 9×10^{16} is c^2. One can therefore calculate the amount of mass that has to be transformed to provide the energy required in accelerating the mass to reach a certain distance. Dividing Equation 3.14 on both sides by c^2 results in this equation

(Equation 3.16) $\dfrac{E}{c^2} = \dfrac{\text{Mass x acceleration x distance}}{9 \times 10^{16}}$

where Mass is in kg, acceleration is in units of metres per second and distance is in metres and is the distance covered during the acceleration.

Numerically it follows that when the product of acceleration x distance divided by 9×10^{16} is greater than one, the energy required to accelerate the mass to achieve that distance is greater than the energy content of the mass itself. But the energy comes from the mass. It therefore follows that in the process of accelerating the mass the mass uses itself up. It extinguishes itself. It also follows that the distance, the extinguishing distance, when this occurs can be calculated. This occurs when $E/c^2 = 1$ The acceleration is given by Hubble's expansion constant, either 4.86×10^{-10} metres per second, or 6.80×10^{-10} if one accepts NASA's figures. This enables the extinguishing distance, or the distance that will be traversed by mass when being accelerated by the Hubble expansion constant before all the mass has been used up in providing the energy required for that acceleration.

The distance therefore is

(Equation 3.17) $$\frac{E}{c^2} = 1 = \frac{1 \times \text{acceleration} \times \text{distance}}{c^2}$$

Extinguishing Distance (metres) $$= \frac{c^2}{\text{acceleration}}$$

$$= \frac{9 \times 10^{16}}{(4.86 \times 10^{-10})}$$

$$= 1.85 \times 10^{26} \text{ metres}$$

$$= 19.56 \text{ billion light years}$$

If one takes NASA's value of the Hubble constant the distance is 14 billion light years.

All those galaxies that accelerated themselves to achieve this distance will have evaporated as energy. There will be nothing visible beyond that distance. It should be noted that these distances are independent of the mass of the galaxy. Any reduction in the mass of the galaxy will be offset by the reduction in energy output per unit time and so the distance remains unchanged. This figure puts a limit on how far telescopes can peer into the distance, and since looking in the distance means looking into the past how far back in time one can see. We will not be able to see the Big Bang. The actual figure is slightly less as some of that mass will also have been used to supply gravitational energy output of the mass.

There is another way at looking at this. This is to determine if there is a time limit as opposed to a distance limit, for the duration of mass. From Newton's laws of motion when starting from rest, distance equals acceleration multiplied by the square of time divided by two.

Applying this to Equation 3.15 and dividing by c squared

(Equation 3.18) $$\frac{E}{c^2} = \frac{Mass \times acceleration^2 \times time^2}{2c^2}$$

But E/c^2 is mass. It cannot exceed the original mass as the energy driving the acceleration is coming from the mass. Therefore when all the mass has converted to energy the value of acceleration squared x time^2/ 2c^2 must equal one. That is

(Equation 3.19) $$1 = \frac{acceleration^2 \times time^2}{2c^2}$$

With a little re-arranging and taking the square root this simplifies to

(Equation 3.20) $$Time = \frac{2^{0.5} \times c}{acceleration}$$

where Time is how long the acceleration continues until all the mass has been used up to provide the energy for that acceleration. Time is in seconds, c is metres per second (3×10^8), acceleration is in metres per second per second (4.86×10^{-10})

(Equation 3.21)   Life expectancy of mass $= 27.66 \times 10^9$ years

If the acceleration constant is $4.86 \times 10\text{-}10$ m/s/s the time becomes 27.66 billion years. If one uses the NASA's Hubble constant figure the time becomes 20 billion years. That is after this time the mass will have extinguished itself because of the need to fuel the constant acceleration over this time. This applies to all masses from a proton to a super giant galaxy. The actual figure will be approximately 10% less than this as some of the mass will have also been used to provide gravitational energy whilst the mass was accelerating. The time, relative to earth time may also be increased if the acceleration produced relativistic velocities and consequent time slowing.

Table 3,1 Data on 80 galaxies and Supernovas

Galaxy or SN title (source)	Velocity v	Apparent Magnitude	k	a	Relativity factor	Corrected Magnitude	Log distance factor	Distance Mparsecs	raw Ho l.yr. billions	Distance l.yr. billions	Time correction factor	Time corrected Ho
1	2	3	4	5	6	7	8	9	10	11	12	13
1 Virgo ©	0.004	*	*	*	8.68E-06	*	*	23.93	50.15	0.08	0.99	49.87
2 Pegasus (b)	0.013	*	*	*	9.17E-05	*	*	70.55	55.28	0.02	1.00	55.19
3 1992al (a)	0.0139	14.6	-0.01	0.13	1.04E-04	14.243	1.789	61.46	67.85	0.20	0.99	66.86
4 Pisces (b)	0.015	*	*	*	1.22E-04	*	*	82.82	54.33	0.27	0.98	53.26
5 Cancer (b)	0.016	*	*	*	1.39E-04	*	*	90.49	53.04	0.30	0.98	51.90
6 1992bo (a)	0.018	15.85	-0.01	0.11	1.76E-04	15.49	2.039	109.34	49.39	0.36	0.97	48.10
7 Perseus (b)	0.018	*	*	*	1.76E-04	*	*	99.69	54.17	0.33	0.98	52.88
8 1992bc (a)	0.0198	15.2	-0.01	0.07	2.14E-04	14.89	1.919	82.91	71.65	0.27	0.98	70.23
9 Coma (b)	0.022	*	*	*	2.63E-04	*	*	122.70	53.79	0.40	0.97	52.22
10 1992P (a)	0.026	16.13	-0.01	0.12	3.67E-04	15.76	2.092	123.57	63.12	0.40	0.97	61.26
11 1992ag (a)	0.026	16.59	-0.01	0.38	3.67E-04	15.96	2.131	135.31	57.65	0.44	0.97	55.79
12 1990O (a)	0.03	16.62	-0.01	0.25	4.89E-04	16.11	2.163	145.49	61.86	0.47	0.97	59.72
13 Hercules (b)	0.034	*	*	*	6.28E-04	*	*	190.18	53.63	0.62	0.95	51.21
14 1992bg (a)	0.035	17.41	0	0.77	6.65E-04	16.39	2.218	165.18	63.57	0.54	0.96	61.07
15 1992bl (a)	0.042	17.31	0.01	0.04	9.58E-04	17.02	2.344	220.79	57.07	0.72	0.95	54.07
16 Pegasus2 (b)	0.043	*	*	*	1.01E-03	*	*	236.20	54.62	0.77	0.94	51.55
17 1992bh (a)	0.044	17.67	0.01	0.1	1.05E-03	17.32	2.403	252.96	52.18	0.82	0.94	49.04
18 1990af (a)	0.049	17.92	0.01	0.16	1.31E-03	17.50	2.440	275.70	53.32	0.90	0.93	49.82
19 1993ag (a)	0.05	18.29	0.01	0.56	1.36E-03	17.47	2.435	271.96	55.15	0.89	0.94	51.58
20 Ursa Major©	0.05	*	*	*	1.36E-03	*	*	306.75	48.90	1.00	0.93	45.33
21 1993O (a)	0.051	17.83	0.01	0.25	1.41E-03	17.32	2.405	254.07	60.22	0.83	0.94	56.58
22 Cluster A (b)	0.052	*	*	*	1.47E-03	*	*	322.09	48.43	1.05	0.92	44.72
23 1992bs (a)	0.061	18.36	0.03	0.09	2.02E-03	18.02	2.545	350.51	52.21	1.14	0.92	47.85
24 Leo (b)	0.065	*	*	*	2.30E-03	*	*	358.90	54.33	1.17	0.91	49.69
25 1993B (a)	0.068	18.68	0.03	0.31	2.52E-03	18.12	2.564	366.69	55.63	1.20	0.91	50.78
26 1992ae (a)	0.072	18.61	0.03	0.15	2.82E-03	18.21	2.582	381.92	56.56	1.25	0.91	51.42
27 Cor Borealis. ©	0.073	*	*	*	2.90E-03	*	*	429.45	51.00	1.40	0.90	45.78

28	1992bp (a)	0.076	18.55	0.04	0.21	3.14E-03	18.10	2.560	363.28	62.76	1.18	0.91	57.34
29	Gemini ©	0.078	*	*	*	3.31E-03	*	*	429.45	54.49	1.40	0.90	48.92
30	1992br (a)	0.084	19.71	0.04	0.12	3.84E-03	19.33	2.806	640.25	39.36	2.09	0.85	33.36
31	1992aq (a)	0.096	19.29	0.05	0.05	5.03E-03	19.00	2.739	548.45	52.51	1.79	0.87	45.66
32	Bootes ©	0.13	*	*	*	9.25E-03	*	*	766.87	50.86	2.50	0.82	41.58
33	Ursa Maj.2 (b)	0.134	*	*	*	9.84E-03	*	*	688.65	58.38	2.43	0.82	48.04
34	1997I (f)	0.157	20.44	-0.33	0.16	1.35E-02	19.64	2.867	737.02	63.91	2.40	0.82	52.70
35	1997N (f)	0.164	20.19	-0.34	0.1	1.48E-02	19.44	2.828	672.72	73.14	2.19	0.84	61.43
36	Hydra ©	0.203	*	*	*	2.28E-02	*	*	1214.72	50.13	3.96	0.71	35.64
37	1997ac (f)	0.271	21.38	-0.55	0.09	4.14E-02	20.39	3.018	1041.41	78.07	3.40	0.75	58.72
38	1994F (f)	0.294	22.08	-0.58	0.11	4.91E-02	21.02	3.144	1393.68	63.29	4.54	0.67	42.30
39	1994 am (f)	0.306	21.82	-0.61	0.1	5.34E-02	20.74	3.088	1225.06	74.94	3.99	0.71	53.09
40	1994I (f)	0.307	21.28	-0.61	0.1	5.34E-02	20.21	2.982	958.75	96.06	3.13	0.77	74.15
41	1997O (f)	0.307	22.97	-0.61	0.09	5.37E-02	21.88	3.317	2073.02	44.43	6.76	0.51	22.51
42	1994 an (f)	0.31	22.14	-0.62	0.21	5.48E-02	20.94	3.127	1340.49	69.38	4.37	0.68	47.25
43	1995ba (f)	0.316	22.08	-0.63	0.06	5.71E-02	21.01	3.143	1388.61	68.27	4.53	0.67	45.71
44	1995aw (f)	0.324	21.75	-0.65	0.12	6.02E-02	20.61	3.061	1151.33	84.42	3.75	0.73	61.29
45	1997am (f)	0.334	21.97	-0.67	0.11	6.42E-02	20.81	3.102	1264.10	79.27	4.12	0.70	55.42
46	1994al (f)	0.337	22.37	-0.68	0.42	6.54E-02	20.89	3.117	1310.08	77.17	4.27	0.69	53.11
47	1994G (f)	0.34	21.52	-1.13	0.03	6.67E-02	19.99	2.938	866.53	117.7	2.82	0.79	93.44
48	1997Q (f)	0.343	22.01	-0.69	0.09	6.79E-02	20.84	3.109	1285.06	80.07	4.19	0.69	55.59
49	1996cn (f)	0.343	22.58	-0.72	0.08	6.79E-02	21.39	3.217	1649.20	62.39	5.38	0.61	37.91
50	1997ai (f)	0.355	22.25	-0.68	0.06	7.31E-02	21.12	3.163	1455.67	73.16	4.75	0.65	47.82
51	1995az (f)	0.355	22.44	-0.71	0.61	7.31E-02	20.73	3.086	1219.65	87.32	3.98	0.71	61.98
52	1996cm (f)	0.355	22.66	-0.71	0.15	7.31E-02	21.40	3.220	1660.33	64.14	5.41	0.60	38.80
53	1995aq (f)	0.357	22.6	-0.71	0.07	7.40E-02	21.42	3.224	1674.79	63.95	5.46	0.60	38.46
54	unnamed (d)	0.36	*	*	*	7.54E-02	*	*	1533.74	70.42	5.00	0.64	44.72
55	1992bi (f)	0.36	22.12	-0.72	0.03	7.54E-02	20.98	3.135	1364.73	79.14	4.45	0.68	53.44
56	1995ar (f)	0.364	22.71	-0.71	0.07	7.72E-02	21.53	3.245	1757.96	62.12	5.73	0.58	36.13
57	1997P (f)	0.368	22.52	-0.72	0.1	7.90E-02	21.30	3.199	1582.48	69.76	5.16	0.62	43.49
58	1995ay (f)	0.373	22.64	-0.72	0.35	8.13E-02	21.17	3.173	1490.28	75.09	4.86	0.65	48.46
59	1996cg (f)	0.379	22.46	-0.72	0.11	8.42E-02	21.22	3.185	1529.42	74.34	4.99	0.64	47.29
60	1996ci (f)	0.382	22.19	-0.71	0.09	8.56E-02	20.98	3.137	1370.76	83.60	4.47	0.67	56.33

61	1995as (f)	0.383	23.02	-0.71	0.07	8.61E-02	21.82	3.304	2015.19	57.02	6.57	0.52	29.68
62	1997H (f)	0.399	22.68	-0.7	0.16	9.41E-02	21.40	3.220	1659.57	72.13	5.41	0.61	43.64
63	1997L (f)	0.412	22.93	-0.69	0.08	1.01E-01	21.73	3.286	1930.31	64.03	6.29	0.54	34.62
64	1996cf (f)	0.423	22.7	-0.68	0.13	1.07E-01	21.46	3.231	1703.12	74.51	5.55	0.59	44.31
65	1997af (f)	0.427	22.96	-0.68	0.09	1.09E-01	21.75	3.290	1949.41	65.71	6.36	0.54	35.23
66	1997F (f)	0.428	22.9	-0.68	0.06	1.10E-01	21.72	3.284	1922.56	66.79	6.27	0.54	36.23
67	1997aj (f)	0.428	22.55	-0.68	0.11	1.10E-01	21.33	3.205	1603.54	80.07	5.23	0.62	49.52
68	1997K (f)	0.434	23.73	-0.69	0.07	1.13E-01	22.51	3.443	2771.86	46.97	9.04	0.34	15.99
69	1997ce3 (e)	0.44	*	*	*	1.17E-01	*	*	1533.74	86.06	5.00	0.64	54.65
70	1997S (f)	0.444	23.03	-0.67	0.11	1.19E-01	21.80	3.300	1994.20	66.79	6.50	0.53	35.10
71	1995ax (f)	0.446	22.53	-0.67	0.11	1.20E-01	21.31	3.201	1588.66	84.22	5.18	0.62	52.38
72	1997J (f)	0.448	23.25	-0.67	0.13	1.22E-01	21.99	3.339	2181.18	61.62	7.11	0.48	29.64
73	1995at (f)	0.465	22.62	-0.66	0.07	1.32E-01	21.43	3.226	1683.74	82.85	5.49	0.60	49.66
74	1996ck (f)	0.466	23.08	-0.66	0.13	1.33E-01	21.82	3.305	2018.12	69.27	6.58	0.52	36.01
75	1997R (f)	0.466	23.28	-0.66	0.11	1.33E-01	22.04	3.348	2229.91	62.69	7.27	0.47	29.43
76	1997cj3 (e)	0.5	*	*	*	1.56E-01	*	*	1533.74	97.80	5.00	0.64	62.11
77	1997G (f)	0.513	23.56	-1.13	0.2	1.66E-01	21.73	3.287	1934.86	79.54	6.31	0.54	42.92
78	1996cl (f)	0.539	23.53	-1.22	0.18	1.86E-01	21.61	3.263	1831.91	88.27	5.97	0.56	49.79
79	1997ap (f)	0.54	23.2	-1.23	0.13	1.87E-01	21.33	3.206	1605.50	100.9	5.23	0.62	62.35
80	1997ck3 (e)	0.97	*	*	*	1.54E+00	*	*	2340.49	124.3	7.63	0.44	55.09

St.Dev Mean +/-1.28

49.10

Column 1. Sources a = Hamuy M et al.,1996, **Astrophysical Journal 112**, 2391.

b = Hoyle F. 1975, in **Astronomy and Cosmology** p81, WH Freeman

c = Hoyle F. 1975 in **Astronomy and Cosmology** p250, WH Freeman

d = Crosswell K. 1992, **New Scientist 136,** 1846.

e = Eisenhammer J & Levaz 1998 **NASA Report** HTTP/ oposite.sstci.edu

f = Perlmutter S et a; 1999 **Astrophysical Journal 517** 565

Column 2. Velocity as a fraction of the velocity of light

Column 4   k a correction factor to obtain total luminosity of the supernova.

Column 5   a , a correction factor also used to derive total luminosity

Column 6   r, a correction factor for the effects of relativity's time dilatation due to the velocity

Column 7 Corrected magnitude, from $(M+k-a-r) \times 0.985$ The 0.985 is due to the absorbtion of 1.5% of the light from the supernova, presumably due to a thin veil of dust surrounding the solar system, a remnant of the supernova that spawned the Solar system. This effect was seen when the predicted Magnitude was compared with the observed (corrected   but without the multiplier) magnitude. The observed magnitude was persistently larger, by an average of 1.5% (range 0.7-4%) of the predicted magnitude. That is there was approximately 1.5% loss of light when that light was in transit to earth. With this correction the time corrected Hubble constant was then comparable to a number of previous estimates of its value..

Column 10   Raw Hubble value, velocity (km/sec)/ Mega Parsec

Column 12   Time correction, that is Age in billions of years less the time taken for the SN light to reach earth/ Age in billions of years.

Column 13 The Hubble value time corrected (see next chapter)

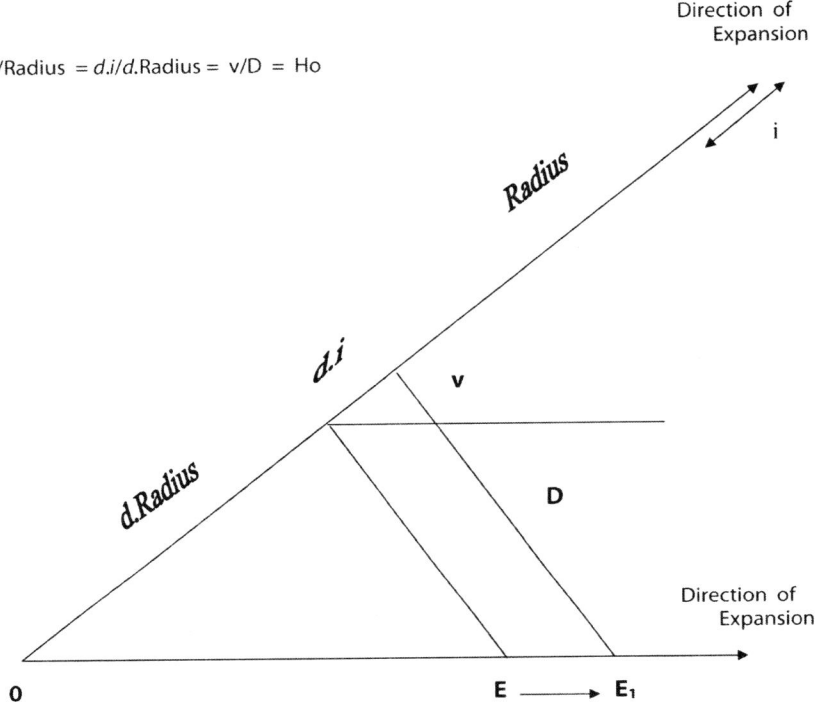

$i/\text{Radius} = d.i/d.\text{Radius} = v/D = Ho$

Figure 3.1 The geometry of the Hubble constant. Or the geometry of an expanding sphere. If *i*, the velocity at the edge of the universe equals c the velocity of light and expansion continues then the ratio v/D must progressively become smaller. This analysis relies on Euclidean geometry where there is no curvatiure of space. This analysis makes no allowance for the transit time for light to travel the distance D. E and E$_1$ are successive distances of Earth from the origin.

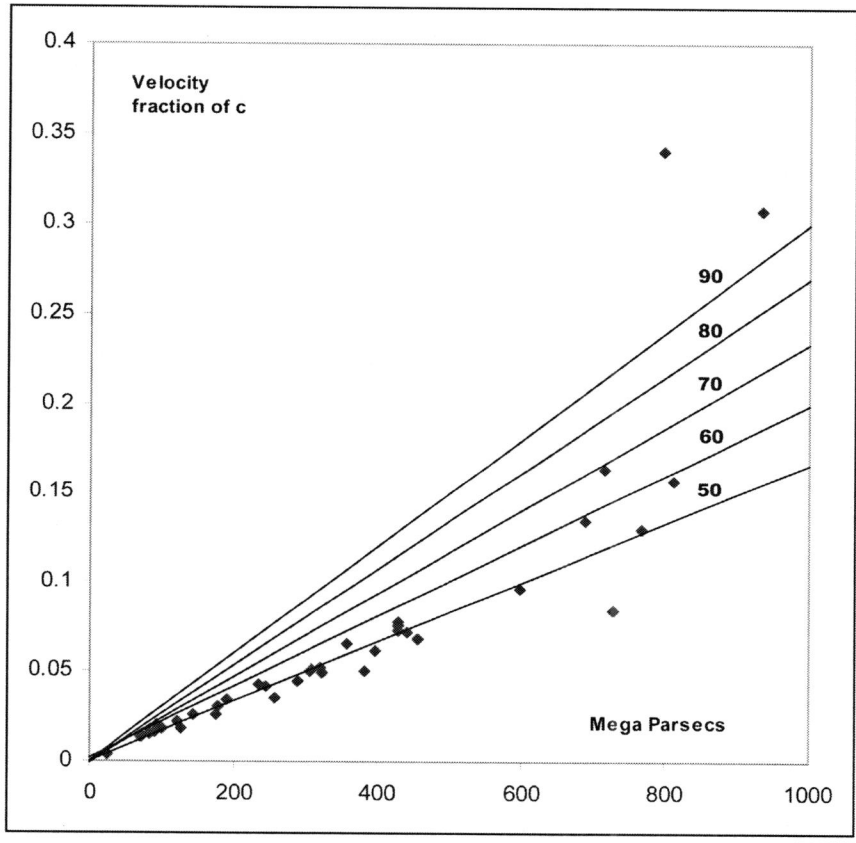

Figure 3.2 The relation between velocity and distance for 38 nearby galaxies and supernova, The oblique straight lines give the distribution expected from five different values for the Hubble constant. The slight rising trend apparent as distance increases was attributed to observer error and the assumption made that the Hubble value is constant throughout space. This should be compared with Figure 3,3 which has data from more distant galaxies and supernovas.

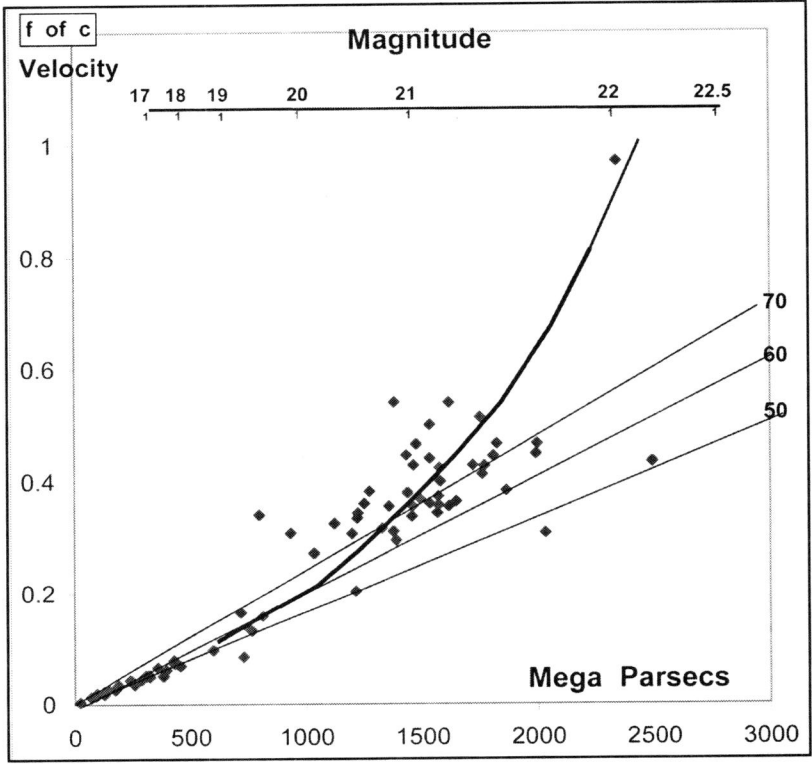

Figure 3.3. The relationship between velocity and distance. The velocity is expressed as a fraction of the velocity of light. The distance is three times greater than in Figure 3,2. The oblique straight lines predict the positions at the three stated Hubble constant values, and assume that the Hubble constant is constant throughout all space. The heavy curved line is the predicted line of the relationship between distance and observed velocity if the Hubble value apparently changes with the age or distance. The magnitude scale gives an indication of the sensitivity of the distance estimates to small errors in magnitude determination at very high magnitudes, and so causes the increasing scatter with increasing distance. The data relate to 80 galaxies and supernovas detailed in table 3,1.

Figure 5

Time adjusted velocity -v- M.parsecs

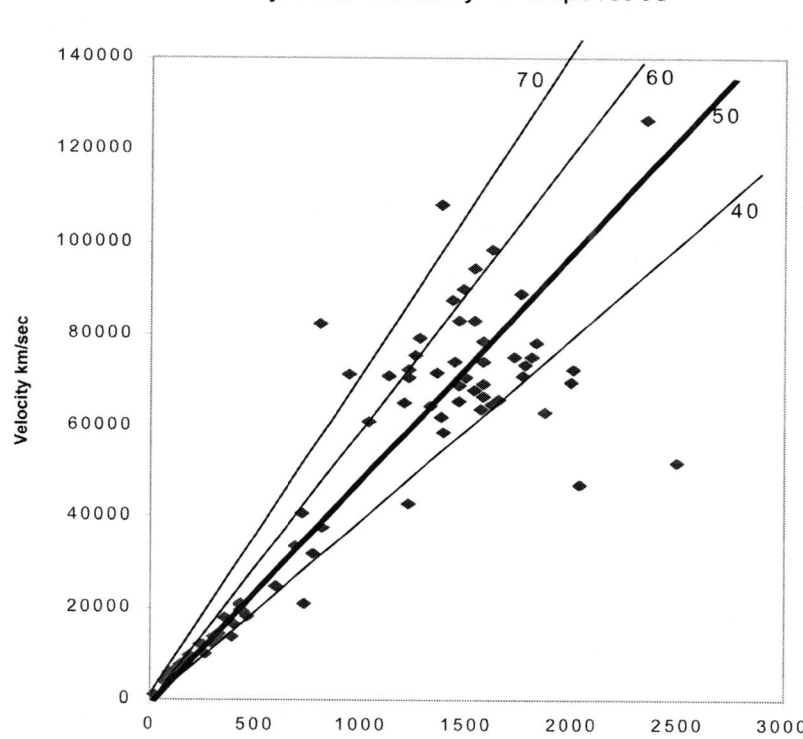

Figure 3.4 The relationship between the time adjusted velocity and distance. The observed velocity was divided by the ratio of the Ages., that is the age of the universe at the time the observed velocity occurred and the present Age, both in Earth time. The numbered straight lines are the forecast relationship at various values for the Hubble constant. The scatter at the time end of the velocity range in due to the hypersensitivity to small errors or changes in brightness in the estimate of distance. The linear relationship between the time adjusted velocity and distance is clear.The range extends to more than half the radius of the

universe, confirming that this constant is defining the fundamental constant of nature, but this is onlty true if time is expanding

Appendix 4. The expansion force mechanism

The expansion force can only accelerate a mass in the direction of its movement if it acts logarithmically. The following shows the principle. The acceleration force arises from and acts on all surfaces of an object, somewhat like a compressive force. If an object is moving is a particular direction, say North, then on the north side of the object there will be opposition to the northward movement. If the velocity is 10 units of distance per second (log value 1) and the force is 2 units (log value 0.3) the potential braking effect should result in a velocity of 0.7 (antilog value 5 units per second). But on the south side of the object the acceleration force will summate with the log of the velocity, resulting in a velocity of log value of 1.2 or 20 units of distance per second. The net value will then be the mean of 20 and 5, resulting in a velocity of 12.5 units of distance per second in the direction of movement. The effect of any force or part of a force, emerging from the object, which is not in line with direction of movement, will be met by an equal and opposite force. Thus sideways deviation from the direction of movement cannot arise from the object itself. It should be appreciated that the proposed 2 units of force is the geometric mean as the energy for this force is being supplied logarithmically because of a time change effect. The latter is described more fully in a subsequent chapter.

However if there is an additional external force this will cause the object to change direction. Then the axis of thrust will align itself with the new direction of movement. It will not require much external force to cause a change in direction. A simple analogy is that of an outboard motor attached to the rudder of a small yacht. It requires a very small force applied to the rudder, small compared with the energy of the outboard motor, to change the direction of movement.

To put this into perspective, galaxies have been observed in clusters and they orbit around each other. If there are two galaxies, each a billion solar mass, separated by a distance of 2 million light years (such as our Galaxy and the Andromeda galaxy), the gravitational force will be

$$g = GM_1M_2/r^2$$
$$= \frac{G(2 \times 10^{30})^2}{(2 \times 10^6 \times 3 \times 10^8 \times 3.15 \times 10^7)^2}$$
$$= 7.5 \times 10^6 \text{ m/s/s of force.}$$

This will produce a net acceleration of 3.7×10^{-16} m/s/s for the billion solar mass galaxy. This in turn would require 140 mega watts which is minute considering that the mass involved is a billion times that of the sun. Despite these tiny amounts of energy the galaxies orbit each other. If the expansion force produces an acceleration of ~5 $\times 10^{-10}$ m/s/s this would require ~10^{18} watts, or ten billion times as much energy as available from the gravity energy from the distant galaxy. Thus the orbital motion is powered by the expansion force but steered by gravity.

There are two consequences to this. The Andromeda galaxy would appear to be heading for a collision with our Galaxy. If the two galaxies were orbiting each other, like two dancers spinning on a dance floor, optically it would appear that Andromeda is heading for a collision because we have no way of determining circumferential velocity of Andromeda's orbit, nor that of our Galaxy. But if our Galaxy is orbiting Andromeda this could account for the observations attributed to the so called Great Attractor, where the latter is a postulated super galaxy hidden on the far side of our galaxy behind the central dusty disc. This postulate accounts for the observational data that our galaxy seems to moving sideways across the universe as well as heading towards the edge of the universe.

It can be calculated that with an orbital radius of 2 million light years and an orbital velocity of one thousandth that of the speed of light (30km/sec) the orbital period would be approximately 13 billion years. This is close to the age of the universe when calculated in earth time. If the red shift effect from the "Great Attractor" shows a velocity much lower than this there is something wrong with the calculation of the age of the universe or that a stable orbit has not yet been achieved.

Chapter 4

A question of time

Defining time, the age of the universe, cosmological time, consequences of time inflation, the globular paradox, speeding up or slowing down

No one knows exactly what time is. We can define and measure it in relation to movement, the swing of a pendulum, the oscillations of a crystal, the passage of the earth around the sun, and so forth. Time is measured in arbitrary units. The choice of such arbitrary units is bizarre since there is a constant, determined by nature, and characteristic of the whole universe now, in the past and in the future. This is the speed of light in a vacuum. It would therefore be logical to standardise time against this universal constant. This is that the second is that period of time for a photon to travel 299,792.4 km in empty space. Such a definition allows elasticity when considering the rate of change of time relative to our existing rate. For we also indirectly use time as a rate fixer, the rate of change of so much per second. This leads to a consideration of the question, is there a rate of change of time itself? It has always been assumed that this rate is constant and always has been constant. There is no proof to support this assumption. According to special relativity theory in the case of a very fast moving space traveller his rate of change of time will be slowed. The velocity of light in the traveller's cabin is normal as far as the traveller is concerned but to the distant observer the velocity of light has slowed. That is there is confusion as to the use of the word. It can refer to a past episode, or the accumulated time, or the unit of time, or the unit of time within a different time frame as in the fast moving traveller's cabin.

It has been propounded that the velocity of light is constant in all time frames. That is it is constant relative to the local pace of time. The relationship between time, velocity and distance is a simple one. The distance achieved is the product of velocity and the duration of time that the velocity is maintained. 10 km/hour for one hour gives a distance of 10 km. It follows that if the distance is unchanged and time slows, for whatever reason, so that, say, the second doubles in length then to an external observer the velocity must halve. But an observer entering into the location where time has slowed will not be aware of this. To him the velocity is unchanged. This must apply to all velocities and not just the velocity of light. But velocity is the result of acceleration. That is all accelerations are constant relative to the time frame. But acceleration is the result of an application of force. It follows that all forces are constant relative to the pace of time. If time slows all forces diminish, including gravity. This has profound significance when considering such exotics as black holes. This is dealt with in a later chapter.

But all time dependant measurements and constants must adhere to the prevailing rate of change of time. These include G, c, velocity, acceleration, the forces responsible for acceleration, the force responsible for the expansion of the universe, down to the oscillations of a crystal in an atomic clock, or frequencies including Doppler shifted frequencies. These Doppler shifted frequencies (the red shift) are imprinted on the spectrum of light. This imprinting must remain the same as that light travels through space. If the observer is in a different time frame he may interpret the velocity according his time frame rather than the time frame prevailing when the Doppler shift actually occurred. But the velocity is relative to that distant time frame. To give a specific example if the Doppler shift shows a velocity that is, say, 0.85c and is arising at a site where the time is twice as fast then the velocity of the emitting object, as far as the observer is concerned, is or should be 1.70c. That is relative to an observer things can move faster than the velocity of light although the observed Doppler shifts will not show this

Time has a direction, the so-called arrow of time. It points in one way only, that is forward. Theoretically we can travel forward in time if we travel fast enough. The theory of relativity shows that we can slow the rate of change of time by moving very fast, and so arrive at a place at some point in the future.

But time travel backwards is impossible, and will also be impossible. We are made up of atoms and molecules and a fundamental block is that no atom can be in two places at the same moment of time. That is if we were to go back in time the atoms and molecules that make up our bodies would have to be in different locations. The carbon atoms that make up the neurones in our brains would have to be scattered among their different origins from the variety of food that we eat. Those atoms would have to be at their original sites. All the theory about wormholes and travelling in loops of time cannot get around this objection. It is often claimed that there is nothing in the laws of physics to prevent time going backwards, that is cause and effect are interchangeable, yet nothing can make the transition from going forwards in time to going backwards in time. Incidentally there is one astronomical hypothesis which makes use of this inter-changeability. . This is the weak anthropic principle. We are here because the constants of nature, the charge of an electron, g, etc. are finely tuned to very precise values, becomes that because we are here those constants of nature must have those precise values. But we did not set those constants so far back in time. That is cause and effect is not interchangeable.

Another problem in attempting to travel backwards in time is at the cross over there would be a moment when the rate of change in time, the pace of time, would be zero. There would be total stasis. Electrons could not move around atomic nuclei. Light would stand still. There would be no force and no heat. Nothing could enter that stasis since by entering there is motion which requires time.

Nevertheless our concept of time is a practical one. It enables things to be accomplished, or causation to be established. It enables us to determine what happened in the past and when it occurred in relation to other happenings that were also in the past. This includes the age of the universe. How long ago was it formed? That particular calculation of the age of the universe rests upon the assumption of an unchanging rate of change of time.

The age of the universe in earth time

Traditionally Time and the radius of the visible universe are presumed to have started with the Big Bang. There must be a relationship between the two, therefore

(Equation 4.1) $$R^n = kT^m$$

where R is the radius of the volume achieved after an accumulated time T, k is a constant defining the relationship, and n is an exponent that could be unity. If the exponent is unity or the two exponents have the same value then radius and time are directly linearly related. The accumulated Time is in earth units of time, where time is presumed to be at a constant rate.

The visible universe is an expanding sphere. Its radius is therefore increasing in length. The equations outlining the theoretical basis of the Hubble constant are given in the previous chapter.

One particular equation enables the age of the universe to be calculated. This is

(Equation 4.2) $$\text{Age} = \frac{D(3v+2)}{3v}$$

where for any galaxy its distance is D in billions of light years and its velocity v is the velocity expressed as a fraction of the velocity of light. This equation was applied to all 80 galaxies and supernovas listed in Chapter 3 and resulted in a mean value of

Age = 13.82 +/- 0.25 billion years of earth time

This figure may be compared with the 13.7 +/- 0.2 billion years that NASA reported. That is there is no statistically significant difference between the two estimates. It should perhaps be noted that if this is accepted then the geometrical analysis given in the previous chapter and equation 4.2 are both valid and therefore the equations which arise from them are also valid, and that includes the equation showing that the Hubble "constant" varies with distance.

The average 13.8 billion years is close to values that have been determined elsewhere. The Einstein-De Sitter model of a "flat" expansion of the universe predicts an age of around 13 billion years. A flat expansion means that the geometry of expansion follows Euclidean lines, in contrast Einstein's General Relativity theory says that the universe is based on a different, a curved geometry, this point is discussed in a later chapter. In an analysis by Gribben and his colleagues of over 1000 nearby galaxies aimed at determining the Hubble constant found that the constant had a value of 52+/-6 km per Megaparsec which was translated to an age of between 13 and 16 billion years. It should perhaps be noticed that a transit time of one billion years, for the light to travel, causes the Hubble "constant" to change by 10% which accounts for the range in his age determination. Thus the simple geometric approach for determining the age of the universe results in figures that match reasonably closely those derived from other sources.

But the figure is wrong. It has to be. It cannot account for the distances to the Sloan galaxies, nor can it account for the variation in the Hubble constant when this is looked at over a very wide range of distances. It cannot account for the anomaly that in order to get to their present locations some galaxies are apparently older than the universe, nor the anomaly that the Globular clusters are apparently older than our Galaxy.

Cosmological time

The assumptions behind the 13.8 billion figure therefore need examining in detail. The key assumption is that the duration of the second has remained constant since the beginning of time. As a consequence extrapolating our present concept of time right back to the beginning results in the 13.8 billion year figure. Extrapolating can lead to error. It is though a very convenient system even if it does not apply to reality. For convenience this system of time measurement will be referred to as earth time. To adhere to it as an article of faith (for there is no objective proof that the second has remained constant since the Big Bang) is very reminiscent of Bishop Wilberforce who maintained as an article of faith that the world was created in 4004 BC. This conclusion was based on his extrapolation of the days as described in the Bible, and was quoted during the celebrated acrimonious Oxford debate that followed the introduction of the theory of evolution in the 19th Century. This same mind set is behind the Creationist's philosophy, that the extrapolation is justified because we believe it to be true. But Science dictates that when the observed data do not fit with established beliefs those beliefs have to be discarded and a new explanation sought. To behave otherwise is to imitate the Creationists.

The Sloan galaxies are more than ten billion light years away and they surround us. The sum of the distances to two such galaxies lying on opposite sides of the universe exceeds the age of the universe. From the site of the Big Bang they must get to their present positions and then shine their star light back to us. The observed data requires that the concept of the constancy of time, which governs the estimate of the age of the universe, must be abandoned.

Although the duration of the second or the pace of time cannot have been constant since the beginning of time, there is some indication that over the last billion years or so, if the pace was not constant, it was close to being constant. This evidence lies in the consistency of the Hubble constant data for reasonably near galaxies, that is those that lie within approximately one billion light years.

The Hubble constant equation can be made constant by multiplying the result by the ratio of ages, that is (Age-D_T)/Age

(Equation 4.3) $$H_o = \frac{652}{(Age-D_T)} \times \frac{(Age-D_T)}{Age}$$

This correction factor was applied to all the Hubble values for the 80 galaxies and Supernova listed in Table 3.1. The results are shown in Figure 4,1. It will be seen that there is almost a horizontal line. The slope of this line is not statistically different from a slope of zero. That is with this correction the Hubble constant is truly constant and has a base value of 49.2 km/second/Mega parsec. Nevertheless this is an extrapolation and may not be true very early in the history of the universe due to interference by gravity. This would be when the universe was small and the consequential gravitational forces were particularly strong.

It was shown in the previous chapter that substituting time for the light to travel the distance of the denominator of the Hubble constant resulted in a unit of acceleration. The value of this unit apparently varied with distance. With this correction factor the unit of acceleration becomes constant. If the base value of the

Hubble constant is 49.2 km/sec/Mega Parsec this acceleration constant has a value of

(Equation 4.4) Acceleration constant = 4.66 x 10^{-10} metres/sec/sec.

For convenience this will be referred to as H_T that is the Hubble constant adjusted for time. That is the expansion constant is constant in all time frames. But the expansion constant is describing a force for expansion. It follows that this must be a fifth fundamental force of nature. It fulfils the criteria for a fundamental force of nature and corresponds to the theoretical prediction described in the last chapter. It can then join the other fundamental forces, the strong and weak intra nuclear forces, the electromagnetic force and gravity. It could be considered an antigravity force although this is not a true reflection since, unlike gravity any distance between different masses does not affect it.

The value of this fifth fundamental force of nature may be compared with the Gravitational constant = 6.67 x 10^{-11} This is the acceleration in metres/second/second of two one kg masses separated by a distance of 1 metre.

That is the acceleration constant is seven times more powerful than gravity. Gravity though increases its strength when the masses are large and the distance between the masses is comparatively small (that is compared with the usual astronomical distances). It is small wonder that the universe was able to expand despite gravity.

The base Hubble constant is $H_0 = 652/\text{Age}$ where Age is the age of the universe in earth time units. But time progresses. In the future when the universe is twice as old as it is now the Hubble constant will apparently have halved. This is due to the expansion of time. The second compared with the present second will be twice as long. When this is allowed for the constant remains constant.

Equally when the universe was half its present age the constant was apparently doubled. That is the period of the second was half its present value. Clocks ticked and pendulums swung twice as fast as they do now. The speed of light was twice as fast as it is now.

When the age of the universe was a quarter of its present value the Hubble constant was four times its value and so on. But it is always constant in relation to its time frame. This time can only be called Cosmological time. Every time the age doubled (in earth time units) the period of the second doubled its length. That is time is slowing in an exponential fashion from being extremely fast to becoming its present earth time value. And this process will continue indefinitely.

It is possible to calculate the length of this doubling period and it is approximately ten billion years (actually 9.76 billion if the age of the universe in earth time is 13.8 billion years). For the mathematically curious this figure is the difference between the two end points—now and half now multiplied by the square root of 2, the resultant is then multiplied by $2^{(n-1)}$ where n is the doubling number counting backwards from the present time. That is to say, from equation 4.1 every ten billion years or so of cosmological time the radius of the universe doubled. There would have been 202 doublings in the size of the radius of the universe if it started as a singularity of diameter one Planck unit, 10^{-35} metres. That is the age of the universe in cosmological time would be around two thousand billions years apparently. But and there is a big but with this figure. Time did not start when the singularity began expanding. This was entirely due to gravity.

When the gravitational acceleration reaches the velocity of light time stands still. Close to the Big Bang the gravitational acceleration would have slowed time very considerably as well as distorted the path photons could take. This effect is examined in more detail in chapter 10.

Technical Note

For various exotic reasons it has been calculated that the mass of the universe has been put at around 10^{80} nucleons—protons and neutrons. Avogadro's number is the number of particles in a gram molecular weight, and has a value of 6.02×10^{23}. From hydrogen, which has a molecular weight of 2.015 grams, it follows that there are 3×10^{23} protons per gram. The sun has a mass of 2×10^{30} kilograms. It therefore has 6×10^{56} nuclear particles. If there are 10^{80} particles in the universe this means that approximately the universe is 10^{24} times the mass of the sun, that is a million, billion, billion times more massive. Thus when the universe was very small all that mass it must would have made the granddaddy of a black hole with a mass of $\sim 2 \times 10^{54}$ kg. When corrected for time slowing this becomes 6.9×10^{59} kg (see Chapter 10)

The radius of this original black hole can be calculated using the standard gravity equation

(Equation 4.5) $\qquad g = GM_1M_2/r^2$

where G is the gravitational constant 6.67×10^{-11}, M_1 is 2×10^{54} kilograms, r is the radius in metres and $g = 3 \times 10^8$ m/s/s. That is g equals c the velocity of light. M_2 is 1 kilogram. Only after the radius had expanded to a radius of at least 70 light years could the gravitational acceleration start to fall below the speed of light. Before then time stood still. For comparison our nearest star is four light years away. With that radius of 70 light years that original black hole was monstrously big. Yet the universe had to have a radius of at least seventy light years before time could start or mass to form and escape to eventually link up to form galaxies. Actually this is a gross under estimate since it ignores the effect of time on mass. This is discussed in the next chapter.

If the universe now has a radius of 13.8 billion light years then since time started the universe has apparently doubled its radius approximately 28 times. If each doubling is worth around ten billion years this puts the age of the universe in cosmological time as ~280 billion years. But this figure needs adjusting for the effects of the expansion of time and so loss of mass, (See chapter 5). The additional mass would create a bigger initial event horizon, well beyond the 70 light years

If the present radius of the universe is 13.8 billion light years (the inference being that it expanded at the velocity of light) its volume is 9.3×10^{78} cubic metres. For various exotic reasons the mean density of the universe has been postulated at about 1 nucleon per cubic metre. There are an estimated 10^{79} to 10^{80} nucleons— the figure varies slightly with different authors, so the inference appears to be justified.

Cosmological time has an exponential form, as opposed to the linear system of earth time. The significant aspect of this system of time is that the various constants, the speed of light, G the gravitational constant and others adhere to this system. This type of exponential change is widespread in nature. The most well known one is the half-life of radioactive materials. It is the exact mirror of the exponential expansion of time. Putting the two together means that radioactive materials decay linearly with time, that is with cosmological time. But there are other examples of exponential change, from the washout of pollutants in a lake, to the growth of populations, to the detonation of explosions or to the progress of a chemical reaction. It also applies to the change in length of the radius of the visible universe.

Consequences of the inflation of time

A number of significant consequences arise from this:

(a) The present rate of change of time is $\sim 1/ (4.3 \times 10^{17})$ where 4.3×10^{17} is the age of the universe in earth time seconds. That is time currently is slowing by this amount every second. As will be seen in the next chapter, this has significant effects on mass and the whole structure of the universe. It governs the whole universe. As will become apparent later it is responsible for our very existence. Without it the earth would not rotate around the sun.

(b) The ratio of radii between different epochs equals the ratio of ages when those radii occurred. This is because of the linear relationship between the radius and time.

(c) Potentially the pace of time when the Big Bang occurred was very fast indeed, although that was slowed by gravity. Once the universe had managed to escape gravity's maw time resumed its natural pace, being very considerably faster than the present rate. Since then time has slowed to become our present pace of time. That is the period or duration of the second has expanded exponentially. The key importance of this is that all the time related things such as the frequency of radiation waves (including the red shift), the rate of change of stellar composition etc. adhere to cosmological time and not earth time. Hence very old stars within our galaxy will appear to be older than the galaxy itself.

(d) It follows from the above that relative to the present time the speed of light then was very much faster than it is now. Energy could therefore radiate to very distant locations without violating the principle of not exceeding the velocity of light. At some point during the travel cooling occurred to allow condensation of energy to matter that would eventually form galaxies but there was plenty of time, that is plenty of cosmological time, for those galaxies to get to their remote positions. A corollary of this is that there would be plenty of time for energy mixing, so that the universe had to become isotropic, that is have in broad terms a uniform number of galaxies per unit volume.

(e) It also follows from this that positing a specific inflation period immediately after the Big Bang is wrong. The original definition of inflation was that the velocity of the radiation of energy exceeded the speed of light in order to achieve the eventual isotropicism. Although the inflation of time hypothesis gives a spurious perspective that the speed of light was exceeded (and hence inflation of volume and isotropicism), it could be argued that what happened was inflation when this is viewed from our present pace of time. But this is playing with words. More significantly in the first seventy years plus (using extrapolated earth time) time did not exist as everything was a black hole consisting of immense amount of energy

(f) It means that the Hubble Constant cannot be constant in earth time units. This is dealt with specifically in a later chapter.

(g) One other fact emerges. The Big Bang was an extremely slow affair. The very slow initial rate of growth would maintain an extremely high temperature, too high for nucleons to form. The universe was very old before it cooled sufficiently for nucleons to form.

The globular clusters paradox

Throughout our Galaxy there are over 150 globular clusters each of which contain thousands of stars. From within any one cluster the spectral composition shows that each member is approximately the same age although there is some variation between different clusters. The clusters are not particularly bright but from a combination of these two factors it was concluded that they are very old. Indeed they appear to be older than 13.8 billion years, the age of the Galaxy (earth time). There has been a more recent satellite observation in which a star was identified as possibly a strayed member from a cluster family. Its brightness suggested that the cluster stars are brighter than first thought and so their dimness as perceived by telescopes is because they are further away than first calculations showed. This in turn suggests that their intrinsic brightness was greater, that is they were burning fuel more quickly and hence though old were not excessively so. Whether such an assumption, arising from a single observation of a stray star that resembles a cluster star can be applied to all the globular clusters given their slightly variable composition must be a matter for debate.

However the expansion of time hypothesis obviates this paradox of members of the galaxy being older than the Galaxy itself. Every time the universe doubled its radius it doubled its age in earth time units. That is when the universe was half its present size and age it was less than seven billion years old (earth time). But each doubling unit takes approximately ten billion years of cosmological time. Natural events, radioactive decay, chemical reactions, etc. adhere to cosmological time. Any stellar changes, which occurred one doubling unit ago, occurred ten billion years ago in cosmological time and the stellar structure and composition would show this. Yet in earth time they would be calculated as being only approximately 7 billion years old. Two doubling units amount to twenty billion years of cosmological time but only ten billion years earth time. And so on. That is using cosmological time the globular cluster paradox disappears. The globular clusters are very old in cosmological time but it is the extrapolation of present earth time, which has generated the apparent paradox

A similar explanation accounts for the very long distances, in excess of ten billion light years of the Sloan galaxies. It gets around the paradox that the sum of the distances between two galaxies that lie on opposite sides of the universe should not exceed the age of the universe. If the age of the universe is calculated in cosmological time this paradox also disappears.

Slowing down or speeding up

A common question is "Is the universe slowing down?" The context usually relates this to the rate of expansion, although it is often not clear as to whether this relates to radial expansion or volume expansion. Surprisingly this question is quite complicated. On the superficial level because time is slowing then the universe is slowing down. The radius is increasing at a constant rate according to earth time, with the rim moving at the speed of light. But according to cosmological time the radius doubles itself every ten billion years, that is it is increasing its rate of expansion.

There is one other effect. As the universe expands so the more distant galaxies are further away from the central parts of the universe. As this distance increases so gravitational drag from the rest of the universe will lessen. The acceleration constant will still be active and the result will be speeding up of the rate of recession of the distant galaxies. This assumes that they have not reached their theoretical absolute speed limit—the speed of light.

The central thesis of this chapter treats time, the second, as increasing that is it is expanding exponentially. That is the speed of light is constant relative to cosmological time as are all the constants of nature. It should perhaps be noted that the speed of light is affected by the medium through which it passes. It is almost half as slow in water. Frequency is not affected. When looking under water the colours remain the same, but colour is a consequence of frequency. This means that any frequency that existed before the velocity changed still persists. If velocity has slowed wavelengths—which are simply fractions of the distances travelled per second—must shorten. If time slows, as during the expansion of the universe, there is no stretching of the waves, but frequency remains constant relative to the pace of time. The red shift remains unchanged relative to the second, as that is an aspect of frequency, not wavelength. A useful analogy is the change in frequency heard when playing a tape recording of music at half speed

Of some interest and possible relevance is that it has been shown experimentally that light comes to an almost stop when in a vacuum above a super cooled atom condensate, that is a collection of atoms cooled to a small fraction above absolute zero. At these temperatures the atoms seem to fuse to form one giant atom. The velocity of light is thus affected by temperature. But the velocity of light defines time. The logical inference is that if temperature influences the velocity of light and the velocity of light in a vacuum defines time, then the cooling that resulted from the expansion of the visible universe is or could be responsible for the slowing of time. When the temperature was so very high the pace of time was also extremely high. This suggestion needs considerable investigation before it can be accepted as being more than an intriguing speculation. Nevertheless both temperature and time are changing exponentially with the change in size of the universe. At near absolute zero temperature time is nearly stopped. Clearly neither can ever reach zero that is where time does change it cannot go to infinity so that is there is no absolute time standstill. Temperature reflects the density of energy. Time would have appeared to have originated out of the energy of the Big Bang although because of gravity it was initially extremely slow. It seems reasonable to conclude that the exponential character of their rate of change is related to the expansion of the visible universe and so dilution of energy. Such a conclusion would mean that there must be an equation linking energy to time, just as there is an equation linking energy and mass. But then applying the hypothetical time energy equation to the mass energy equation it becomes apparent that time, mass and distance are all inter related.

This is also obvious when one examines the Planck units and of what they are composed. At the time of the singularity, assuming that there was a singularity, all that existed was in Planck units of time, distance and mass (that is energy) were all variations of the same fundamental units.

There is no reason to object to the principle that gravity slows time by itself. It is not necessary to postulate that gravity induces a trampoline-like depression of space as has been suggested in general relativity theory to account for the observed slowing of the passage of photons as they pass very close to massive objects like the sun. A simple slowing of time will suffice, that is gravity expands or slows time.

It has been suggested that photons are massless although they have momentum. They are simply packets of energy. But so is mass according to special relativity theory. That is just as a satellite will deviate from its path when passing close to a massive body, the sling shot effect used so effectively in the directing of space probes, so gravity would cause the photon to deviate from its path. The deviation is very slight due to the momentum of the photon. That is gravity appears to have a dual effect, it can change the direction of movement of a photon as well as slow time. See Chapter 6.

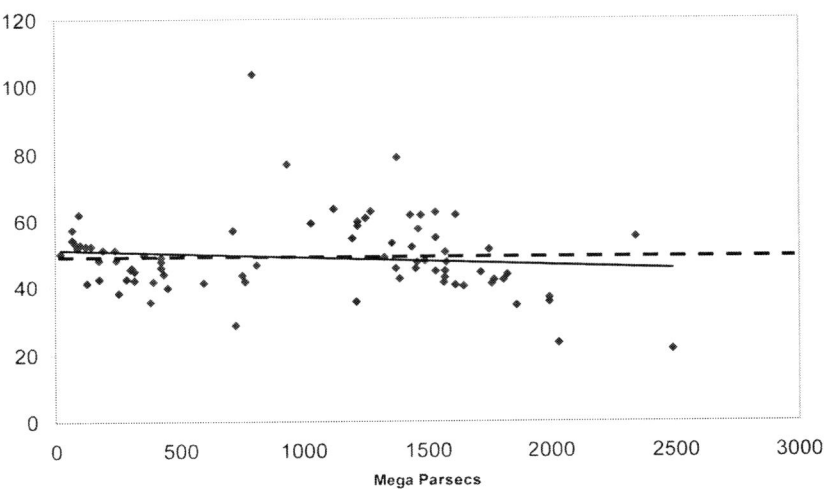

Figure 4.1 The corrected Hubble constant expressed as a function of distance. The Hubble constant derived from each observation of 80 galaxies or supernovas had been corrected by the ratio (Age-D_T)/Age where Age is 13.8 billion years and D_T is the transit time for light to travel the Distance, D. Although the regression line shows a slight downward slope this slope is statistically not different from zero. (dashed line). It results in a base Hubble constant of 49.2 km/sec/Mega Parsec, or as an acceleration constant 4.66 m/s/s

Chapter 5

Time, Mass and Gravity

The Pendulum paradox, mass and velocity, the pendulum and time, mass time relationship, the solar thermostat, a new source of energy, the work of gravity and its fuel source

Since the days of Galileo Galilei there has been fascination about the pendulum. It was the most accurate method of determining time for several hundred years until supplanted by the atomic clock and then by the measurement of oscillations of a crystal. It was quickly found that the period of the pendulum oscillation was inversely related to the square root of the gravity's acceleration, g. The equation describing this mathematical relationship for a pendulum of length l is

(Equation 5,1) $t = 2\pi(l/g)^{0.5}$

But g is the acceleration resulting when two masses M_1 and M_2 are separated by a distance, r, the acceleration falling in accordance with the inverse square law. It is dependent on mass. The equation describing this is

(Equation 5,2) $$g = GM_1M_2/r^2$$

where G is the Gravitational Constant with a value of 6.67×10^{-11}. Put simply if two masses of 1 kg each where separated by a distance of one metre, the resulting acceleration due to the gravitational attraction between the two is 6.67×10^{-11} metres per sec, per sec. The two masses would take a long time to get together but eventually would.

Putting the two equations together and assuming the two masses are equal (or one mass is exactly one kg) this simplifying results in the following equation.

(Equation 5,3) $$T = \frac{2\pi r}{M} \times \frac{l^{0.5}}{G^{0.5}}$$

Since G is the Gravitational constant and so is constant in all time frames, whilst l and r are fixed distances which are independent of time, it follows that Time is related to the inverse of mass. If time should expand, that is the second become twice as long, the mass would halve. But mass is energy, according to Einstein. Therefore the reduction of mass should be accompanied by a release of energy. The amount of energy released in turn is very substantial, being 9×10^{16} Joules per kilogram of mass that has been shed. (from Einstein's famous equation, 9×10^{16} is merely c^2).

The Planck equation describing the quanta of energy available for any given frequency also confirms that time slowing must cause a reduction or shedding of mass. This is elaborated in more detail in Chapter 6.

Equation 5,3 also means that just as the speed of light is constant in all time frames so is the inverse relationship between mass and time preserved in all time frames. This has crucial importance, as will be seen when discussing solar energy control. We in fact owe our existence to this simple fact. Mass in any particular time frame has its own size. Changing to a different time frame means that mass must either be shed as energy or mass must be acquired by absorbing energy, or if desperate by atomic fusion.

It was noted (see chapter 10) that since time started time has expanded to double itself 18 times. That is the second is now approximately 250,000 times longer than when time started? That is the mass of every proton and neutron would have been 250,000 times greater than at present if they could have existed. But it was too hot. Every doubling of time indicates a doubling in the radius of the universe. It also indicates that the volume increased eight-fold with each doubling of the radius or time unit. Temperature changes linearly with volume. That is Temperature changes 8 fold with each doubling unit. The temperature of space when time first started was 8^{18}, or ~2×10^{16} times higher than the present temperature of intergalactic space. (Although this is currently assumed to be ~2.3°K this cannot be true of intergalactic space, as 8^{18} represents an increase in temperature to ~2×10^{16}. This aspect is dealt with in Chapter 8). Even if the temperature of intergalactic space is 10^{-8}K then at ~18 doubling events ago the starting temperature would have been more than two hundred million degrees. Protons cannot exist at this temperature. Their velocity would be above the speed of light (the velocity of hydrogen atoms changes at a rate of ~3.2 metres/second/degree Celsius, (See technical note) and is even faster as the velocity approaches relativistic speeds. It was not until time and so volume, had expanded so that the temperature would be low enough for the velocity of protons to be below the speed of light that protons could exist. The temperature then would have been ~10^7 degrees Kelvin. This occurred only seven doubling units ago This carries with it the inference that the universe as we know it, consisting of atoms, stars and galaxies etc is seven doubling units old, with each doubling unit worth about ten billion years of cosmological time. It is ~seventy billion years old in cosmological time. Translating that into earth time units the universe was approximately 1-2 million years old since time began before atoms could form. The radius of the universe then would have been almost forty thousand light years, smaller than our Galaxy (radius fifty thousand light years). Protons would have been almost 130 times more massive than they are now. This undoubtedly favoured rapid gravitational condensation to form galaxies.

Since mass is now 1/130 as much as it was when it first formed the question arises as to where did all that energy go. A significant fraction went to accelerate the multi-billion stars that are in each of the fifteen billion galaxies in an expanding universe. To accelerate a mass to half the speed of light (and many galaxies have velocities in excess of this) standard equations of the energy of acceleration show this requires the energy equivalent to about 14% of its mass. Approximately 2% went on supplying gravity (the acceleration constant is about seven times greater than the Gravitational constant). The rest must have been in the form of heat or light.

But there is a problem. The popular version of relativity theory states that relative to a stationary observer a moving object increases its mass. The equation describing this is

(Equation 5.4) $$M = m/(1-v^2)^{0.5}$$

where *v* is the velocity expressed as a fraction of the velocity of light. This equation has exactly the same form as Einstein's relativity equation describing the change of time with velocity

(Equation 5,5) $$T = t(1/(1-v^2))^{0.5}$$

Yet clearly the inter-relation based on equation 5,4 cannot be right. A star, or black hole, or an atom on crossing the event horizon of another black hole would be travelling at the speed of light. It therefore should have infinite mass. Even before then it would have been accelerating at huge speeds apparently generating great mass. That in turn would have had a powerful gravitational effect and would wreck the spiral structure of a galaxy. This is not what is observed.

Another example is the deceleration of very fast moving protons. If this interpretation of relativity theory were right then decelerating the fast moving protons would cause their masses to revert to more normal size. That should be accompanied by a great burst of energy. Instead when decelerating the protons cool very rapidly and there is great instability of the proton beam, that is the mass of the individual protons are rapidly changing with each proton struggling to get to a normal size. A similar situation was noted in the early experiments trying to obtain nuclear fusion. The magnetic mirror reflectors of the earlier attempts to make a reactor for controlled thermonuclear fusion did not show great bursts of energy as the beams approached the mirrors to be slowed down and then reflected backwards.

Examination of the origins of Equation 5,4 show that it is based on what happens when a moving mass is passed to a stationary observer. It is on receipt of the mass that the stationary observer senses that the mass has increased. That is the mass is also stationary. In contrast equation 5,5, the time equation starts with a stationary clock being accelerated to high velocity and the fast moving clock is sending messages back about its time keeping.

Labelling equations 5,4 and 5,5 shows the effect

(Equation 5,6) $$M_{stationary} = m_{moving}(1/(1-v^2))^{0.5}$$

(Equation 5,7) $$T_{moving} = t_{stationary}(1/(1-v^2))^{0.5}$$

Rearranging equation 5,6 so that it is in the same shape as equation 5,7 results in

(Equation 5,8) $$m_{moving} = M_{stationary}(1-v^2)^{0.5}$$

A moving object experiences time slowing and so loses mass. The relativity statement therefore should be that relative to a mass held by a stationary observer a moving object decreases its mass. A more profound analysis is that the quantity of mass is relative to its time frame, although the relationship is an inverse one. The observer plays no part in the relationship, he just happens to be in the particular time frame. This follows the pattern that the velocity of light is constant relative to its time frame.

Yet there are laboratory experiments which have been interpreted as showing that moving protons increase their mass. The essence of these experiments lie in the observation that deflecting a fast moving proton (e.g. applying a magnetic field at right angles to the direction of movement) through a set angle requires greater and greater magnetic energy the faster the protons are moving. Einstein is reported to have said, "That is what I expected". But again this result needs to be examined in detail. Part of the effect undoubtedly is because the faster moving protons would have been exposed to the sideways force for a shorter time. This can easily allowed for and still there is a discrepancy. Another factor is that the protons will be experiencing time slowing so that to the protons the exposure time for the sideways force is even less. Again that can be allowed for and some discrepancy still remains. However the third contributor is time itself. The force particles on reaching the protons will themselves slow down. This is obvious if the particles were travelling at close to the speed of light. The force particles cannot go faster than the speed of light that prevails in the environment they are in. But force is mass times acceleration. The denominator for acceleration is time squared. That is if the time of the environment, that is the time experienced by the fast moving particle, expands by a factor of two the force is reduced to a quarter of what it would be in normal time. The end result is with the fast moving protons although their mass is reduced in linear proportion to the expansion of time; the force acting on them to push for a change of direction is reduced by the square of the expansion of time. It therefore is less effective. Therefore to achieve the desired change in direction of movement of the fast moving protons the energy of the magnetic field has to be cranked up. It is this that has been wrongly attributed to an apparent increase in mass. It is also this effect which explains why nuclear physicists find it necessary to have such large storage rings for their fast moving protons. Smaller rings, with a higher rate of change of angle per unit length of the circumference, require considerably more additional power for their deflecting magnets.

There is one other aspect of the relationship between time mass and gravity of interest. An atomic clock placed at the equator would run slow by 2×10^{-7} seconds per year compared with an identical clock at the pole. This is because of the relative velocity at the equator due to the rotation of the earth. This is a simple variation of the famous twin paradox of one twin rocketing off into space at high velocity and on returning to find himself younger than his twin. It is of interest that Einstein in his 1905 paper was aware of this and specifically stated that the clocks at the Pole and equator should be balance wheel clocks to show the time difference. He subsequently wrote a note saying "in contrast to a pendulum clock". He would seem to have been aware of the paradox posed by the interpretation being placed on his mass velocity equation. If velocity at the equator increased mass the gravitational force should be stronger; the pendulum clock would run faster. Whereas if time had slowed the pendulum clock should run slower.

Thus the apparent paradox posed by asking the question how fast would a pendulum on the moon swing if the moon orbited the earth at high velocity (chapter 1) can now be answered. The moon's mass would be reduced (not increased, as relativists would have it); gravity would be less the period of pendulum would increase, that is it would show time being slowed. In addition because of the high velocity time would also be slowed and the pendulum swinging more slowly would reflect this. The rules of physics have not been breached. This again is in accord with relativity's fundamental axiom. This is that the rules of physics are the same for all observers.

The consequences of mass being lost when time slows are profound. The energy released is responsible for fuelling the expansion of the universe, and for fuelling the work done by gravity. It is also responsible for the energy output of the stars. It provided conditions that enabled life to evolve.

When mass is shed it must release energy. The faster the rate of shedding the more the intense the energy released. From the previous chapter Time is currently slowing at a rate of 2.43×10^{-18} per second, that is the energy content of this release per atom is extremely low. Since the energy release is carried by force particles, e.g., photons and gravitons etc., the energy density of each force particle would be very low. Such a description would apply to a graviton. Quantum theory states that the total amount on energy in a packet or quantum of energy depends upon frequency. The higher the frequency the more is the energy. Conversely when the energy is low the wave frequency is low. Since the wave is travelling at the speed of light this means that the wavelength must be very long. Gravity waves must therefore have very long wavelengths.

When the rate of mass loss of a nucleon is much higher the energy content of the energy carrying particle would be much higher and could spread into radio waves initially, then through infra red and so on up to and including gamma radiation. Such a scenario obviously occurred with the detonation of the original fission nuclear bomb. The rate of release of mass was determined by the temperature of the fission which in turn controlled the velocity of the atoms which slowed their time frame down causing more mass to be shed both as energy particles and sub atomic particles and so increasing the release of energy. This occurred in progressively greater amounts faster and faster ending up in a catastrophic explosion of energy with much of it of very short wavelength.

Thus the form of energy produced will depend upon the rate at which time is slowed. But if the pace of time is reduced by only an almost infinitesimally small amount some of it will appear as additional gravitational energy. Once the fast moving body has reached a stable velocity the gravitational force generated will then reflect not only the reduced mass but also the expansion of time All forces are constant in all time frames. If that force crosses to a faster time frame its effect will be accentuated by the ratio of the square of the two time frames i.e., $(t_1/t_2)^2$.

The other effect of changing time and mass is of fundamental importance. Indeed if it were not for this effect life on earth could not exist. This is the solar thermostat.

The solar thermostat

One of the remarkable things about the universe is that individual stars are able to maintain a constant temperature for billions of years. The surface temperature may vary between different stars as it depends upon the intrinsic mass of the individual star but stars of the same type have the same colour (which denotes temperature). The distribution of the mix of sizes of stars in other galaxies is similar to that of our Galaxy. Yet we see them when they were more than many millions of years younger than the present age of the stars in our galaxy. Since their colour denotes their temperature then the temperatures have been the same for all these years. In fact our sun is one of a very common lot. Each star is consuming, that is fusing many hundreds of tons of hydrogen every second. The end product of this fusion is helium. The mystery is despite fusing so much mass why does not the system develop into a runaway chain reaction resulting in the explosion of the star. Yet it does just this when the hydrogen supply becomes too low.

It has always been assumed that the control mechanism was simply the balance between radiation pressure, which would make the sun expand and gravity, which would otherwise cause the sun to contract. If the sun got too hot it would expand and so cool when gravity would return it to its normal size. Such a system operates for the variable stars like the Cepheids. But for a big mass that is the sun, that system is too crude to account for the remarkable thermal stability of stars like the sun. There has to be some mechanism, fundamentally simple, and one that would emerge automatically from the properties of hydrogen. That control mechanism must limit the enormous energy that results from hydrogen fusion from precipitating a violent chain reaction.

The control is clearly dependent upon the amount of available hydrogen. When the hydrogen becomes too depleted there is an increase in temperature which blows off the outer skin as the red giant whilst the core of the star tries to become a helium burning star. It does not always succeed. The mass margin is narrow as too much mass allows overheating with eventual detonation of a type 2 supernova. The stability of the hydrogen fusion system as compared with the helium fusion system is that the temperature for hydrogen fusion is very significantly lower than that for helium fusion, so allowing a margin for any transient temperature overshoot. For helium fusion that range is more limited. For a helium burning star any significant overshoot and there is entry into the next sequence of fusion be it to form carbon, oxygen or any other light element with their even higher temperatures and a rapid progression to a type 2 supernova.

Individual hydrogen burning stars do vary in their surface temperature according to the mass of the star but this is a reflection of the amount of fusion going on within the star and the thickness of its overlying blanket of unused hydrogen. The increase in mass would cause an increase in pressure, which would compress the central core and shorten the path length between the atoms, so increasing the probability of a fusion event for any set of hydrogen atoms. The increased mass would also have more gravitational energy, so that there would be, at any one time, more of the hydrogen being accelerated to relativistic speeds. The consequence of this is a faster rate of fusion and so a faster rate of fuel consumption but the temperature of the fusion process is the same so that there is no escalating chain reaction.

Solar ignition is presumed to have started with gravitational compression and heating. This excites the hydrogen molecules and they move very quickly. Because they are moving quickly their time slows, that is their seconds become longer. This causes a reduction in mass which is shed as energy. This raises the temperature even further and this in turn increases the velocity of the hydrogen atoms (by now it is too hot for molecules) so that they form a hot plasma of fast moving protons. This creates a positive feedback raising the temperature even further. This increases even more the velocity of the protons resulting in yet more mass shedding. Something must give. This is achieved by shedding the energy as solar radiation.

For fusion to occur four atoms of hydrogen must join, either all at once or through a series of steps, deuterium, tritium etc. that has been called the carbon cycle (as carbon catalyses the reaction). Helium has an atomic mass that is 4.0026. Its four nucleons average a mass of 1.00065. It originates from four hydrogen atoms which together were slightly more massive. Each hydrogen atom has a mass of 1.00794 and each atom has to lose 0.00729 of this mass. Applying this to Equation 5.8, the mass velocity equation, allows the required velocity to be determined.

(Equation 5.9) $\qquad 1.00065 = 1.00794(1-v^2)^{0.5}$

The resulting temperature in the middle of the sun is ~10 million degrees (See technical note). But perhaps slightly lower

At collision the atoms are at a transient standstill. They must absorb all that lost energy to regain their mass or else fuse. Fusion relieves the protons of their energy debt. More importantly it diverts the excess energy from precipitating a chain reaction.

The thermostat is surprisingly simple. The temperature determines the velocity of the atoms. If it over heats, the protons move too quickly and so shed too much mass as energy. On collision to standstill there is still an energy debt to be paid despite the amelioration of some of the debt by fusion and so the colliding protons must absorb energy, thereby lowering the temperature.

If there is under heating there is less mass shed but the collision velocities are still high enough for the inertia of momentum to overcome any charge repulsion and so fusion occurs. Some mass has still been shed but there is now an excess amount of energy from the fusion itself and so the temperature rises. The system is self sustaining as long as the pressure and therefore the path length between protons (and so the probability of collision) are maintained. Although some of the energy is radiated away which would allow the system to cool the temperature remains hot enough for fast enough velocities so that the system warms up again. Temperature control is obviously relevant to any system of sustained controlled hydrogen fusion.

It should be noted that there is at the moment no coherent explanation of the precise control of the sun's temperature. Yet ice core research has shown that over the long term the temperature of the sun has remained remarkably constant not withstanding the occasional blip due to transient ice ages. A fluctuation of only fifty degrees in the temperature of the sun would cause such wild oscillations of the temperature on earth that life as we know it could not have evolved. As it was the various Ice ages, infrequent as they were, contributed to the extinction of many species.

Another effect of the time slowing effect on mass applies to the earth itself. A substantial part of that energy will be as heat and this must be responsible for heating the earth's core. The current explanation that it is due to the decay of radioactive materials is not sustainable on quantitative grounds. This is elaborated upon in more detail on page 215

The work of gravity

Isaac Newton's laws of motion state that a moving object will maintain its movement and direction of movement unless acted upon by a force. The application of a force so that movement occurs means work is being done. Energy is being expended, where energy is defined as the ability to do work At the equator because of the revolving earth the sea is moving at a rate of approximately 500 metres per second relative to a stationary observer in space. It should continue at this velocity in a straight line, but is deflected by a force called gravity such that in 12 hours it is moving at the same speed but in the opposite direction. The mass of the sea is very substantial and so this constant changing of direction consumes a huge amount of energy. That energy has to be replaced. But it is continuously being expended which requires a continued source of replacement.

But the work does not stop there. The earth has the same pattern. In six months as it circulates around the sun, it is moving at the same speed but in the opposite direction. This same pattern applies to all the planets. That is even more work, requiring huge amounts of energy. All this gravitational energy comes from the sun.

But it does not stop there. The solar system is in a revolving galaxy so that over a period of a billion years or so it will have changed direction to move in the opposite direction. This applies to all the billion plus stars in the Galaxy. And it does not stop there. This scenario is repeated in all the galaxies in the visible universe.

There is only one possible source for this vast amount of energy. It must come from the loss of mass, where each kilogram lost provides 9×10^{16} Joules.

The speeding interplanetary probes can have their velocities increased by the slingshot effect as they pass around a massive object, the sun, Venus, Jupiter etc. But increasing the velocity of an object requires energy. This is taken from the gravitational field of the planet or sun in question. This requires the transmission of energy. Even the humble moon, whose gravity causes the tides, transmits energy so that during the 25 hours of the earth's rotation under the moon's gaze, a huge body of water is lifted up by more than a meter against the pull of earth's gravity. But as the earth turns away from the moon this mass is released only for an adjacent mass of water to be raised. There is also the counter balance on the opposite side of the earth of work being done to lift the water and so minimise precession of the globe.

Light and gravity have much in common. Both describe the transmission of energy, be it photons or gravitons, but photons have difficulty in moving the smallest of masses, although they can move atoms within a molecule as in photosynthesis, whereas gravity can move whole galaxies. Both obey the inverse square law. Both arise from masses. For photons it is the jump of electrons from one orbital state to another around an atom. For gravity, where the energy per square meter is much greater it must come from the nucleus of the atom itself. Since gravity is continuous the nucleus of the atom must be losing mass continuously.

The expansion of time provides the clue. Time and mass are inversely related. Time is slowing, or expanding, at a rate of $\sim 2.43 \times 10^{-18}$ of a second per second. As the second expands so mass must be lost. Losing mass means releasing energy. Since all the atoms in the universe are shedding mass this releases a vast amount of energy, more than sufficient to cover the huge demands made upon it because galaxies, stars, and planets move in orbits. As will be seen in the next chapter not all the shed mass is used to provide gravitational energy.. Only a fraction of it is gravitational energy. Indeed although the energy consumed by gravity throughout the universe justifies the word astronomical it pales into insignificance compared with the energy required for the expansion of the visible universe. This expansion is on going. It uses a lot of energy. This must be replaced. The only source of energy on this scale is from the release of energy when mass disintegrates.

There remains a mystery. Why do galaxies collide? Our Galaxy and our neighbourhood twin Andromeda are thought by some to be due to collide in a couple of billions years or so, that is if they are not orbiting each other. The two galaxies were formed when the universe was much younger and smaller. But they were closer together then. They originated from the same spot, a microdot, according to the Big Bang theory. They should have virtually the same trajectories. There is the expansion force, but this is constant (see next chapter) and still continues. Although the expansion constant is greater than the gravitational constant because gravitational force depends upon the distance between two bodies, gravitational force is greater for bodies that are comparatively close together.

The alternative explanation as to why the two galaxies have not collided already is that they did not arise from the same nearby location. If the universe was already extremely large before protons could form (See chapter 10) then the fragments of the great cloud of protons that went on to form galaxies could well have arisen some distance from each other. The resultant trajectories would then send them on a collision course. Otherwise the mystery remains, why do galaxies collide so late in the lifetime of the universe?

In summary therefore the cosmological slowing or expansion of time causes mass to be reduced. This releases energy and is the fuel source of gravitational energy.

There is one other aspect of the work of gravity which is somewhat surprising. In the analysis of the solar thermostat it was stated that the solar thermal process started with gravitational heating causing the hydrogen nuclei to increase their velocity and this in turn caused mass to be shed as energy. Calculation, using the standard special relativity equation, has shown that the energy required to increase the velocity of a proton exactly equals the energy shed from the lessening mass. There is no apparent net gain of energy. Obviously when collision and fusion occurs the energy debt is settled, but fundamentally this means that the energy from the entire solar radiative output is supplied by the energy released by time slowing.

The results of this calculation are profound as far as obtaining energy from hydrogen fusion. It means that the energy required to heat up the hydrogen so that the atoms move fast enough for fusion exactly equals the energy that would be obtained from the fusion process. Controlled hydrogen fusion as an energy source of the future seems destined to fail. But there is a caveat. This relies on the sun's thermostat effect when it is getting too cold. This is the inertia effect of momentum to overcome charge repulsion at fusion. Then there would be a gain of energy output. But the energy would has to be removed quickly otherwise the excess energy would merely heat the incoming cooler hydrogen nuclei. If these nuclei were too cold their velocity and so momentum energy may not be enough to cross the charge repulsion barrier. The isotopes of hydrogen, deuterium or even tritium have greater mass and therefore greater momentum relative to the charge repulsion and so would achieve fusion at lower temperatures with a greater margin of reliability of positive energy output. The amount of energy required to achieve a set effective temperature would be slightly higher as their velocity change per degree Centigrade would be less than that it is for hydrogen

Technical notes

1. Calculating the internal temperature of the sun

The speed of sound in air is independent of pressure, at least at pressures close to ambient. It is dependent upon temperature. At 0°C it is 332 metres per sec. Thereafter it rises by 0.6m/s/degree Celsius. The velocity depends upon the mass of the elements. For the thermal energy input using the standard energy mass velocity equation, the equation becomes

$$\text{Energy} = \frac{M_1 \, (v_1)^2}{2} = \frac{M_2 \, (v_2)^2}{2}$$

where M_1 is the molecular weight of gas 1 and v_1 is the velocity in metres per second of the molecules in gas 1., and so on for M_2. Air has an average molecular weight of 28.8, hydrogen has a molecular weight of ~2. Ignoring the starting value of 332 as too small when dealing with very high temperatures the relative velocities per degree Celsius are therefore

Air 0.6 m/s/degree Celsius

Hydrogen molecules 2.27 m/s/degree Celsius

Hydrogen atoms 3.2m/s/ degree Celsius

That is at ten million degrees the velocity is above the velocity of light. But as they approach this velocity the hydrogen atoms lose mass, making them even faster at these extreme temperatures

In the sun for hydrogen fusion to helium each atom of hydrogen must lose 0.7% of its mass, i.e. change from 1.008 to 1.001. The change of mass with velocity is given by the equation

$$M = M_{initial} (1-v^2)^{0.5}$$

$$1.001 = 1.008(1-v^2)^{0.5}$$

where v is the velocity expressed as a fraction of the velocity of light. This change in mass is produced by a velocity that is just under 12% of the velocity of light. If the change in velocity of hydrogen atoms per unit temperature remains unchanged this means that the temperature is just less than eleven million degrees. As the increased velocity resulting from the loss of mass approaches this high temperature the rate of change of velocity per degree is likely to rise. This would lower the sun's internal temperature to ~ten million degrees. It could be lower if the velocity of random motion of atoms with temperature is further affected by the very high pressures that must occur in the depths of the sun.

2. The relation between time and energy

If time is defined by the pendulum equation then energy and time are inversely related in some way. The pendulum equation is

(Equation 5,12) $$T = 2\pi (l/g)^{0.5}$$

But $g = GM_1 m_2/r^2$ where r is the distance between the two masses M_1 and m_2, G is the gravitational constant and l is the length of the pendulum Therefore

(Equation 5,13) $$T = \frac{2\pi (l.r^2)^{0.5}}{(GM_1 m_2)^{0.5}}$$

Substituting energy, E, for mass then

(Equation 5,14) $$T = \frac{2\pi r (lc^2 c^2)^{0.5}}{(G.E.E)^{0.5}}$$

$$T = \frac{2\pi r c^2}{E} \times \frac{l^{0.5}}{G^{0.5}}$$

Since c has a denominator of time and G has a denominator of time squared energy is inversely related to the square of time. Energy itself is a force of acceleration. Time is inversely related to the square root of energy. Dimensionally the equation shows the numerator contains length cubed, that is volume. It follows that as the energy density decreases so time would expand. The possibility exists therefore that it is the expansion of the universe that is responsible for the slowing of time

Chapter 6

Solar gravity and time

Solar gravity and its energy cost, gravitons, relativity theory and gravity, solar eclipse data analysis and interpretation, the refraction of light by expanded time.

The acceleration due to the sun's gravity, experienced by every kilogram of a distant body such as a solar system planet, can be calculated from the following equation

(Equation 6.1) $$g = GM/r^2$$

Where G is the gravitational constant, M is the mass of the sun (2×10^{30} kg) and r is the distance to that distant body (in metres). Acceleration multiplied by mass is force.

The gravitational force exerted by the sun on the planet Mercury (mass 3.13×10^{23} kg) is just over 10^{22} Newtons. The same force is exerted by the planet Mercury on the sun. The energy per second that is transferred is vastly different. It is the force multiplied by half the square of the acceleration. The gravitational acceleration from the sun felt on Mercury is approximately 0.04 metres per second per second. This yields a net transfer of energy that has a mass equivalence of over two and half thousand kilograms per second. In contrast the gravitational acceleration on the sun from Mercury is 6×10^{-9} metres per second per second which yields energy transfer in mass equivalents of 0.4 grams. Mercury's gravitational force has virtually no effect on the sun. This sort of calculation has been used to determine approximately the mass of planets that orbit distant stars. It has the limitation of not being able to detect small mass planets like Mercury or Earth, as their gravitational forces would have so little effect on the star.

The sun's gravitational domain extends at least as far as the Oort Clouds. Within that domain are many astronomical bodies, comets, meteors, asteroids, planets, each orbiting the sun in its own fashion. Without the sun's gravity these bodies would continue in a straight line flying away from the sun. But the sun's gravity causes all these bodies to alter their courses continuously so that they each follow an orbital path. But all this adjusting of the courses by these bodies consumes a lot of energy, which must come from the sun. The energy expended on the planets can be calculated.

Energy per second is force multiplied by the distance moved in that second as a result of that force. The distance achieved in one second is half the value of the acceleration. It follows that, for the earth, the force expended by the sun's gravity to move the earth is

(Equation 6.2)   Energy = Mass x acceleration x distance

$$= 5.8 \times 10^{24} \times 5.9 \times 10^{-3} \times 5.9 \times 10^{-3}/2$$

$$= 3 \times 10^{20} \text{ Joules per second}$$

Mass equivalent $= 3.33 \times 10^{3}$ kg i.e Energy $/c^2$

127

In contrast the total electromagnetic radiant energy received by the earth from the sun has a mass equivalence of only 2 kg/second. That is the gravitational energy supplied to the earth is more than 1500 times the amount of heat and light energy received from the sun.

The total amount of the sun's gravitational energy used just to move the planets can be calculated. It is the energy equivalent of approximately 7.7 thousand kilograms of mass per second. See Table 6.1

The sun's total radiant energy output can also be calculated using the earth's solar constant. At a distance of 500 light seconds from the sun the earth receives. 1400 Joules of radiant energy per square metre per second. This is the solar radiant energy constant. A hypothetical shell or sphere with radius of 500 light seconds has a surface area of $4 \times \pi \times 500^2 \times c^2/3$ square metres. This means that the total radiant energy output per second from the sun, expressed as kilograms of mass, that is dividing the Joules by c^2, is

$$1400 \times 4/3 \times \pi \times 500^2 \; Kg$$

The sun's total radiant energy output amounts to $\sim 3 \times 10^8$ kilograms of mass per second. (the similarity of this figure to the speed of light is purely coincidental). That is since the sun first formed it has expended the equivalent of nearly 10 Earth masses in radiant energy.

Although a gravitational constant can be determined for each planet, to compare this with the radiant energy constant is somewhat spurious. The gravitational energy depends upon the mass of the planet whereas the radiant energy depends upon the cross sectional area of the planet at its equator. Nevertheless some insight can be obtained as to the enormous amount of energy the sun emits as gravitational energy. The planet Mercury receives slightly less than 2 kg/sec as radiant energy but over a thousand times as much as gravitational energy.

A tentative indication is that at a minimum therefore the sun's total energy output as gravitational energy is the energy equivalent of six thousand billion kilograms of mass per second.

If the projection of the moon's gravitational energy for the tides on earth is correct then half the moon's gravitational energy passes through the earth, unabsorbed. Since Mercury and earth have similar densities it is a reasonable assumption that only half of the solar gravitational energy reaching Mercury is utilised. The inference is that the total amount of mass lost from the sun to provide all its gravitational energy output is at least ten thousand billion kilograms of matter per second. Over a billion-year period, the sun would have used up a mass equal to a thousand times the mass of the mass of the earth simply providing gravitational energy.

Of interest the amount of gravitational energy that could be expected from the sun as a result of time slowing of the universe, has a mass equivalence of only 273 kg/sec. That is additional gravitational energy is continuously being generated, or rather released. The only possible source is the energy released when mass is destroyed. This in turn is the result of the high internal temperature of the sun with its associated very high velocity of the hot atoms. Their time slowing causes a loss of mass. Some of the energy from that shed mass goes to form gravitational energy. This is a continuous process fuelling the sun's continuous energy expenditure by adding additional gravitational energy to the background gravitational energy supply produced by the sun's mass.

Gravitons

The current thought is that gravitational energy is transmitted as gravitons. These are packets of energy similar to photons. Gravitons have a much longer wavelength and their energy content per graviton is very much less. The relationship between the wavelength and the quantum of energy each wave carries is given by the equation

(Equation 6.3)   Energy $= h \times$ frequency

where h is Planck's constant and frequency is in Hertz. Since frequency is c, the velocity of light divided by wavelength, a long wavelength implies a much smaller quantum of energy per wave.

A side issue is what happens when an electromagnetic wave is travelling through a medium where the velocity of light is slower, e.g., a light beam travelling through water. The wavelength is reduced but the frequency remains the same, that is the energy content is unchanged. Hence there is no change in colour perception (which depends upon frequency) when viewing things underwater, even though the velocity of light in water is very substantially slower than the equivalent in air or space.

The question then arises as to what happens with time slowing. Energy is mass times c^2 but c is constant relative to its time frame. Similarly Planck's constant is also a fundamental constant that is constant in all time frames. Planck's constant has a denominator of sec². A slower time frame means that both c and Planck's h are less relative to normal time.

(Equation 6.4)   Energy $= Mc^2 = h \times$ frequency

Frequency is also constant relative to the time frame. Thus the velocity of light is much reduced in water, yet photosynthesis continues using green chlorophyll in the chloroplasts. If both h and frequency are reduced with time slowing it follows that the quantum of energy of a graviton reduces with time slowing. Since energy is a very dilute form of mass it follows that mass must be reduced when time slows, that is the second expands. This provides a proof that mass and time are inversely related. Other proofs are given in the technical note

Since this energy is responsible for the gravitational force this analysis provides confirmation that force is constant relative to the time frame, and by inference all accelerations and therefore all velocities are constant relative to their time frames.

By virtue of the very long wavelength of gravitons, a wavelength measured in kilometres, the small quantity of energy means that the shape of the graviton is an extremely long and very thin wave. The nearest analogy is an extremely thin javelin. Some gravitons must therefore be able to pass though the interstices of mass, between its molecules and so penetrate deep into mass and even pass through it. There is a similarity with neutrinos, but the latter must have an even smaller cross sectional area. That is the gravitons must be able to reach those molecules at the very rear of the mass.

The equation describing the gravitational force between masses must mean that every nucleon in the receiving mass absorbs the same number of gravitons. Atoms vary in their size e.g., iron has 56 nucleons, uranium has 238 nucleons, all packed tightly together in the atom's nucleus. When considering those molecules with a very large number of nucleons, it means that some of the gravitons must pass through the bulk of the nucleus of the atom to reach those nucleons at the rear of the nucleus. It follows that nuclei of atoms have a large amount of empty space.

Relativity theory and gravity

From the equation of special relativity (see technical note) time is related gravity.

(Equation 6,5) $Time_{(slowed)}^2 = c^2/(c^2-g^2)$

The end result is a dimensionless ratio describing the enlargement or expansion of time relative to normal time. It follows that when g is very large the denominator becomes smaller, that is time slows or expands. When g equals the velocity of light time becomes infinitely slowed, that is in effect time ceases.

What is critical about this equation, derived purely from the special relativity equation, is that there is no mention of space.

General relativity theory proposes that close by a body of substantial mass time and space become conjoined in a continuum resulting in the production of gravity. But if the special relativity equation is true then space has no dimensions. It is an empty nothingness, just a void. But the general relativity theory proposes that space or rather space-time is curved when close to a massive body. For anything that is curved it must have dimensions and a direction of curvature. Furthermore a three dimensional block that becomes curved must have a volume and the volume of room that this block occupies must increase the greater the curvature, leaving gaps against the next block that is further away from the massive object. But there cannot be gaps. Therefore space must stretch on one side, the side furthest away from the mass, and compress on the side nearest the massive body. That is space is a material with physical properties. The gravitational force is supposed to emerge out of the changing shape. The nearest analogy of the stretching of space is the surface of a trampoline that is stretched and depressed by a heavy mass. The depression is supposed to occur in another dimension.

This still does not resolve the mathematical conflict between the two relativity equations when considering time and gravity. The special relativity equations are based on Euclidean geometry. The General Relativity equations are based on Reimann geometry—geometry of curved lines. If General Relativity theory is true and applies to the universe then the Special relativity equations are false. Among those equations is the famous $E=mc^2$

Another difficulty arises with the space-time concept if it is held that space- time is curved. A number of galaxies show jets of light arising from slowly collapsing black holes The longest jets seen in these galaxies extend for distances almost half the radius of the galaxy, that is approaching 25 thousand light years, a billion, billion kilometres. Despite being in the heart of galaxies with the galaxy's consequent gravitational field they show no sign of any curvature. That is any curvature must be extremely slight even over this distance. Nevertheless the equations of General Relativity predict that a beam of light traces a curve when close to a massive object, with the curvature decreasing the further away from the massive object. The solar eclipse data apparently show a curvature measured in arc seconds when the light from a distant star traverses the sun's gravitational field. But that traverse is only a few lights seconds long. Even if the gravitational field within a galaxy is very weak there should be some curvature when the light beam travels a very substantial distance within that field. This does not fit with the observations of these jets. They are very very straight.

A particular question does arise. The light from very far distant galaxies and supernovas, arising when the pace of time was much faster, has been passing through progressively zones of slower and slower time. That is the velocity of those photons has become progressively slower, relative to their starting velocity whilst en route to Earth. Could this have caused that light to be refracted? That is do the photons travel in curved geodesics rather than in straight lines? It will be recalled that there is no refraction of light when the direction of the light beam is at right angles to the zone when the velocity of light is slower. Such a situation applies to the light from those distant galaxies. Those photons would all be meeting the progressively time slowed universe at right angles, irrespective of the origins of the photons. The conclusion is that light from those distant galaxies does not travel in geodesics. The distance travelled by those photons represents to linear distance to the light-emitting galaxy.

But the biggest problem of all as far as the General Relativity theory is concern is it does not account for the continued supply of large quantities of gravitational energy and how that energy to be transported across space. The nearest home example of this transport of energy is the energy of the tides. The tidal energy from the moon is very substantial

The question then arises are the mathematical conclusions of General Relativity theory true. The answer is yes But the more important question is do they apply to reality? Historically the example of the epicycles show the problems that can occur when applying what were perfectly good mathematical conclusions, based on very limited evidence.

General relativity is a beautiful theoretical construct with marked aesthetic properties, much like Picasso or Dali paintings. These too have marked aesthetic qualities. Like the paintings the general relativity construct may not describe the real universe but it is true to its own internal logic.

Three predictions were made based upon general relativity theory. The first was the red shift of fast moving receding stars and galaxies. But this was quickly ruled out. The red shift was fully explained by the Doppler effect.

The second was the orbit of the planet Mercury. Repeated observations of this orbit indicated that it did not seem to follow strictly Euclidean geometry lines. There were slight delays in its appearance at certain times. In particular the perihelion, the point where the planet was nearest to the sun (the orbit is elliptical) appears to be gradually orbiting the sun. Curved space resolved these difficulties. In curved space the light from the planet would be curved.

The third prediction was that light beam from a distant star passing close to a large mass would be curved. Solar eclipses would be the proof. The stars would be apparently displaced and displaced by a predictable amount if the sun were close to the beam's path. Subsequent studies of the change of angle on viewing distant stars when their light passed close to the sun during a solar eclipse were cited as providing observational proof of the general relativity theory. It was firmly believed that in space the beam of light maintains its position relative to the block of space through which it is passing. If the block is curved the beam will be curved. The theory allowed that magnitude of the change of angle to be calculated. In fact the observational data obtained during a solar eclipse was that the change was greater than expected by up to 15% (See The Golem, H.Collins and T. Pinch. Cambridge University Press 1993). Nevertheless there was an apparent curvature. Equally the observations about the orbit of Mercury had been used to justify General Relativity. Relativity proved a good predictor for the path of the planet and its timetable. It was held that these observations could only be explained if the light followed a curved path. Therefore the conclusion was that the universe followed Reimann geometry and so General Relativity theory applied to the universe. With that the concept of space-time continuum was developed.

There are still nagging difficulties. If space curved around a large mass then as that mass moved, such as the sun's orbit around our Galaxy, space should react. Another problem is how does a photon know when to adjust its trajectory to curve.

The refraction of light by slowed time.

But there is another possible explanation for the curvature. It relies on what is a very common place observation. This is that when a beam of light enters or leaves somewhere where the velocity of light is slowed, the light is refracted.

The extent of the refraction depends up on the angle of incidence of the light beam on the refracting material and the wavelength of light and the velocity of light in the refracting medium. This occurs when a beam of light passes through water, transparent oil, glass, and diamond, in fact anything, which sparkles and produces a rainbow spectrum. It is caused by the wave property of light. As the foot of the wave enters the slow time zone before the crest of the wave it causes the wave to swivel. It will swivel the other way when leaving the slow time zone as the crest of the wave is now travelling slower than the foot.

The analogy is a supermarket trolley when one of its forward wheels catches on something and turns more slowly. Another analogy is the sudden jerk of the steering wheel of a moving car when one side of the car runs into a deep puddle of water. Unless corrected by the driver the car will change direction.

Clearly when a beam of light passes close to the sun it encounters a gravitational field which is slowing time. It is therefore refracted towards the sun. In doing so it passes into zones of stronger and stronger gravity, and so is refracted more. The whole beam becomes increasingly curved. By the time the beam passes the sun's equator it is at its closest to the surface of the sun. As the beam continues it starts to pass out of the sun's gravitational field and is refracted in the opposite direction. During its transit through the gravitational field the photons in the beam will be experiencing time slowing so that the photons emerge late out of the sun's strong powerful field. This particular effect is very marginal. The beam also will emerge at a slightly different angle from the angle of incidence. This is due in part to the sun's own movement across the light of sight. (The sun orbits the Milky Way galaxy and though the orbital period is large, perhaps millions of years, yet the orbital path is so long that the sun is travelling at a significant number of kilometres per second. It is this movement which is partly responsible for the ellipsoid nature of the different planetary orbits.) That is the path the beam of light traces as it approaches the solar equator is not mirror imaged as it leaves the sun. The angle of incidence and the curvature of the gravitational field are more significant factors. The effect is reminiscent of the refraction of light seen when light shines through a spray of water droplets. The actual changes in angle seen during an eclipse were averaged at 1.98 arc seconds, whereas the prediction based on general relativity was 1.70 arc seconds. Calculation shows that gravitationally induced time slowing has an extremely slight effect in causing the apparent path of the stars light during a solar eclipse.

The major contributor to the refraction is the sun's corona. This is an envelope of rarefied gas that surrounds the sun. The inner part of the corona is seen during a solar eclipse. It is kept in position against the sun's gravity by the radiant heat from the solar surface where the temperature is $6,000°K$. It is an like an atmosphere. Atmospheres refract light. This refraction is why, on earth, the sky appears blue. Starlight passing through the outer parts of the corona will inevitably be refracted slightly.

There are two consequences of the refraction explanation. One is that the change of angle will vary depending upon the angle of incidence and so how far the beam is from the surface of the sun when at its closest. The second is that it predicts that there will be a difference in apparent angle change between those stars whose beams pass in front of the sun's direction of travel compared with those stars whose light beam passes aft of the sun. This is so even though the beams may be equidistant from the sun. This is all testable. Curiously relativity theory makes no allowance between those stars that appear to lie fore as compared with those that appear to lie aft of the sun.

Thus the interpretation of the equations of general relativity, that space is curved may be wrong. It is the light beam, not space, which is curved, but less by gravity than by simple refraction Space therefore remains as an empty nothingness, a boundless void. It is emphasised that this is just an alternative explanation for the changes in stellar angle that were seen during an eclipse of the sun.

There is though the possibility of verifying the refraction concept. It depends upon the angular resolution of radio telescopes. In general with refraction the longer the wave length the greater is the refraction angle. Radio waves have a much longer wavelength than light waves. It follows that if radio waves when passing close to the sun are refracted more than 5 arc seconds this will be proof positive of the concept. A pulsar or distant radio galaxy that becomes positioned behind the sun periodically consequent upon the earth orbit around the sun could provide the test. At the moment as far as the author is aware radio telescopes do not have a sufficiently fine resolution

The other "proof" of the General Relativity hypothesis is the odd behaviour of the orbit of the planet Mercury. The ellipsoid orbit itself is gradually rotating around the sun. But this could be due to the strange recently discovered force, currently labelled gravito-magnetism. This is an acceleration force whose properties are still under investigation. The possibility exists that this could be a manifestation of Burhhard Heim's second undiscovered fundamental force of nature.

Another part of the evidence in support for General Relativity's description of curved space are the gravitational lens. These are duplicated and occasionally triplicated images of very distant galaxies. The images are comparatively close together with nothing luminous showing in between. It has been postulated that these are examples of curved space around some very massive object. But the numbers do not add up. The distance between the images suggest that they are at least a galaxy diameter apart and probably much more. But the gravitational force at the edge of a galaxy is remarkably slight. The gravitational force from a thousand suns only ten light years distance is 1.5×10^{-11} m/s/s. This figure is far too small to have any effect on the light path (otherwise there would be obvious gross changes in the light parts of the starlight passing close to the surface of the moon So one is left with something that is causing refraction.

A large cloud of hydrogen gas, say a sphere twice the diameter of a galaxy i.e., 100,000 light years, if it has an average density of a billion hydrogen molecules per cubic metre would still amount to a mass of a handful of suns. That is the gravitational pull from such a comparatively tenuous mass is too weak to form a galaxy. At its thickest though it could absorb light from a distant galaxy. Near its edges it could easily refract light slightly, just as the upper reaches of our atmosphere refract light. That is the so called gravitational lens could well be patches of very thin clouds are too weak to form a galaxy on their own and too far from a proper galaxy to become incorporated with it..

But if the General Relativity theory in practice is not applicable to the universe then the whole concept of a space-time continuum disappears. Time is independent of space and space is an empty dimensionless void with no physical properties. And E will still equal mc². And it is wrong to describe straight lines as curves.

The interaction between time and gravity has a very dramatic effect when close to a black hole. A black hole by definition has a surface gravitational acceleration that equals the velocity of light. A solar mass star spiralling towards a black hole when at a distance of ten light seconds from the surface of the black hole will experience a time slowing of 0.01%. That is it will have lost 0.01% of its mass as energy When it gets to one light second's distance its time will have expanded to 2%, causing a loss of mass of 2%, or several thousand earth masses as energy. At 0.1 light second's distance time will have expanded by 8%. Another 27,000 kilometres closer, when it reaches 0.01 light second distance, time will have expanded to over 9%.

The loss of mass consequent upon this time slowing will be increased energy production. This will increase the infalling star's orbital speed from the release of pent-up acceleration energy as well as the gravitational energy. This increase in velocity will slow time further. The increased orbital speed will also slow down the descent of the infalling star. It can be calculated that if the infalling star was sun sized and it took a hundred years to descend from 30,000 kilometres to 3,000 kilometres from the surface of the black hole, the rate of mass loss and so energy release would be 4×10^{18} kg per second. This can be compared with the tentative suggestion that the sun is losing mass (to provide gravitational energy as well as radiant energy) at a rate of 6×10^{12} kg per second.

That is if the same ratio of gravitational energy to radiant energy persists the infalling star would be almost a million times brighter than our sun and would persist with this brightness for a hundred years.

There are three other factors that need consideration. The first is that as the orbital velocity of the infalling star increases its red shift of its own hydrogen fusion cycle and the parent galaxy's recessional velocity would apparently increase. This could give rise to the impression that the star and its associated black hole are much further away that they really are. That is they may be not quite so bright as currently thought. The second consideration is that the increased orbital velocity would slow time further and so accelerate the rate of mass loss and energy production. The third is that the greatly increased gravitational energy release from the loss of mass from the very fast orbiting infalling star would cause the black hole to spin. The effect is analogous to repeated whipping of a spinning top.

What has been described is the production of a quasar.

Table 6.1 Gravitational energy and the planets

	Mercury	Venus	Earth	Mars	Jupiter	Saturn	Uranus	Neptune	Units
Mass	0.313	4.72	5.80	0.62	1840	551	84.2	101	$Kgx10^{24}$
Radius	2.44	6.05	6.38	3.38	71.4	60.4	25.9	24.7	$M x10^6$
Volume	0.61	9.27	10.9	1.62	1520	923	72.7	63.1	km^3x10^{20}
Density	5.15	5.10	5.34	3.83	1.21	0.60	1.16	1.60	kg/m^3
Distance	1.93	3.61	4.99	7.63	25.9	47.7	95.7	150	Lightsecs
"shell"	0.468	1.63	3.12	7.25	84.5	285	1150	4860	$x10^6c^2$
X-area	0.187	1.15	1.28	0.36	160	115	21.1	19.2	$x10^{14}m^2$
Solar accn	39.4	11.3	5.9	2.54	0.214	0.065	0.016	0.0065	m/s/s
Energy(g)	2420	3000	1010	20	437	11.5	0.108	0.0216	$x10^{17}J$
Energy(G)	2690	3333	1120	22.2	486	12.8	0.12	0.024	kg
Energy (r)	1.95	3.440	1.99	0.24	9.24	1.96	0.089	0.033	kg
Ratio	1380	971	564	92	52.6	6.5	1.34	0.723	(G/Er)

Solar gravitational acceleration data from g=GMsun /r2

Energy(g) is gravitational energy from the sun to the planet per second from

$$\text{Energy (g)} = 0.5 \times M \times \text{acceleration}^2 \text{ in Joules}$$

$$\text{Energy (G)} = 0.5 \times M \times \text{acceleration}^2 /c^2 \text{ in kg}$$

E(r) is electromagnetic or radiant energy received from the sun in kg/sec.

The Shell area is the surface area of a hypothetical sphere enclosing the sun. The sphere has a radius equal to the distance between the planet and the sun. It is in units of 9×10^{16} square metres or c^2. It defines the area through which the sun's radiation energy must pass at this radius distance.

Technical Note 1

Proofs that time and mass are inversely related

Proof 1 The pendulum equation

$$T= 2\, \pi\, (l/g)^{0.5}$$

(Equation 6,6) $$= 2\,0\, \pi r(l/G)^{0.5}/M$$

where G is the gravitational constant, l the length of the pendulum, r is the distance from the centre of gravity of a 1 Kg mass that experiences the gravitational force being applied to it. M is also the square root of the mathematical product of two different masses which are combining to produce the Gravitational force. For a pendulum on earth M is the mass of the earth. Mass and time are inversely related

Proof 2. From Quantum theory

The energy of radiation is given by the equation

(Equation 6,7) $$E = h \times \lambda$$

where h is Planck's constant and λ is the frequency of the radiation. This is the number, n, of waves or packets "quanta" of energy being emitted per second. Each wave contains one quantum of energy, where the quantum is the minimum amount of energy that can exist independently.

$$E = Mc^2 = h \times \lambda$$

$$M = h \times \lambda/c^2$$

$$\lambda = n/t$$

(Equation 6,8) $$= hn/c^2t$$

That is Mass and time are inversely related

Proof 3. From Special Relativity theory

$$T_{expanded} = t_{normal}(c^2/(c^2 - v^2)^{0.5}$$

(Equation 6,9) $$= t_{normal}(c^2/(c^2 - g^2t^2)^{0.5}$$

This can be re arranged to give

(Equation 6,10) $$g = (c/t)((T^2 - t^2)/T^2)^{0.5}$$

 and just as in the pendulum analysis, when g and t are inversely related so also must be time and mass.

Technical Note 2

A relationship between quantum theory and Relativity

This has hitherto been difficult to establish. But from the above equations

$$M = (r^2c/Gt)) \times ((T^2-t^2)/T^2)^{0.5}$$

$$= hn/c^2t$$

It follows that in a static situation where there is no expansion of time

$$r^2c/G = hn/c^2$$

$$r^2/G = hn/c^3$$

That is gravity and quantum theory are also related

It also follows that if time and mass are inversely related, and mass is a form of energy which exists in quantum units, then the expansion of time must be by the addition of quantum units of time. This has to be distinguished from the passage of time, that is the progress from one moment to another. It follows that the quantum unit of time has the form $v^2/(c^2 - v^2)$ where v is the minimum possible velocity.

Chapter 7

Time and the Cosmic Menagerie

The effect of time on galaxies, stars, white dwarves, neutron stars and pulsars, black holes, quasars, Jupiter, the moon and earth, comets and asteroids

The universe consists of a large number of objects of interest that make up the cosmic menagerie. They can be divided into groups. The first are the objects with significant mass. There are the galaxies with their content of main sequence stars, that is stars that follow a particular sequence, being brighter, bluer and hotter with increasing mass. In addition there are neutron stars and pulsars, black holes and quasars. Objects of smaller mass in this set are the planets, asteroids, comets, meteors, and dust clouds.

Next come the exotica, red giants, white dwarves, brown dwarf stars, MACHOs and WIMPS, respectively massive and weak objects. Whether they all exist is open to argument.

The second group is the ephemera, supernovae of various types, gamma bursters, virtual electrons and virtual positrons. Following on are the rays, radio waves, infra red, light, ultra violet rays, X-rays, gamma rays, cosmic rays, gravitational rays. Finally come the particles, neutrons, protons, electrons, quarks of various colours and orientation and their antiparticles. Then there are the mesons and neutrinos of various types, Z particles and the elusive Higgs boson etc, together with the particles of energy, the photons, gluons and gravitons that together make up the particle zoo.

It is of interest to examine the effect of time change on the various large masses that form the major part of the cosmic menagerie.

Galaxies: Galaxies come in various sizes and shapes. These are the ellipticals, the spirals and the irregulars. A large spiral galaxy such as our Galaxy can contain several billion stars but galaxies can be as small as a few million stars. A moderately common shape is the barred spiral which appears to be two incomplete spirals joined by a bridge or bridges. Presumably these are either large spirals breaking into two as a result of each half containing a large black hole, or else the fusion of two galaxies that are meeting edge on.

Why there is such a sharp morphological division between the flat spirals and the ellipticals is due to factors occurring early in the formation of the galaxy. A galaxy starts as a condensation of a vast cloud of a mixture of hydrogen and helium. Gravity causes the cloud to condense. As it condenses the middle of the great cloud becomes very dense. Patches condense forms a large number of giant proto-stars, akin to the globular clusters that bestrew our Galaxy. Their very size causes them to move towards each other jostling and fusing. Some of the jostling is tangential causing a certain amount of spinning. Slowly this central mass gathers together under gravity whilst the spinning continues. But this spinning mass at the centre is key. As it contracts under its own gravity the spin rate increases, just as a spinning skater spins up when the arms are withdrawn to the sides. The smaller the volume the faster the spin.

If this mass is large enough, say with a radius of approximately ten light minutes, then its circumference at the equator is about one light hour. A rotation of one turn per hour would cause atoms on the equator to be travelling at near the speed of light. For comparison Jupiter has a rotation period of ten hours. Atoms travelling at relativistic speeds suffer time slowing and loss of mass. The shed mass becomes energy. The equator then becomes exceedingly hot and bright and releases excessive amounts of gravitational energy. The result is a substantial difference between the gravitational pull at the equator compared with the pole. This difference has the effect of pulling all the would be stars in the proto-galaxy into a single equatorial plane. Their gravitational drag would also cause the entire galaxy to start spinning. The effect of this would be to slow the rotational rate of the central mass and so a cooling of the equator. But the formation of the flattened spiral is primarily due to the effects of time slowing on mass, that time slowing being, in turn, due to the rotation rate of that central mass.

But the central mass will still be condensing to form a giant star. This will quickly exhaust its fuel reserves. It would explode as what can only be described as the granddaddy of all supernovas, a truly giant supernova. It would leave behind a massively dense clinker, which would cool, gravitationally collapse and form a big black hole. If the sun were to form a black hole its radius would be 666 kilometres. A million solar mass black holes would have a radius 2.2 light seconds, virtually the same radius as our sun. Although this giant mass would be physically relatively small, its gravitational strength would be monstrous. The energy released by the giant supernova would push out the stars near the centre of the galaxy creating a bulge on the galactic disc surrounded by a halo of dust. Some the nearest stars would eventually be dragged back to the centre by the gravitational pull of the black hole. The end result is a spiral galaxy as we see it. The Sombrero galaxy is a typical result. Galaxies whose central mass have little to no spinning will end up as elliptical galaxies or as the smaller irregular galaxies.

At the centre of a standard spiral galaxy is a massively dense black hole with the gravitational equivalent of up to a million solar masses and perhaps more. It would seem that this large gravitational nucleus is necessary to bring an amorphous but large bunch of stars into the typical discoid shape of the spiral galaxies. It follows that all spiral galaxies, including our own, have at their centres such a massively dense black hole.

Because galaxies are composed of fuel consuming stars they must have a definite shelf life. The commonest star is a G star with a life of about fifteen billion years. The estimate is that the half the mass of any galaxy is composed of visible stars, the rest is dust, brown dwarves and billowing clouds of hydrogen, smaller black holes and spent white dwarfs etc (ignoring the so-called mystical dark matter). These stars will die out either as ending their lives as supernovae; or else as burnt out wrecks as failed helium stars. Supernovae explosions are not very efficient at destroying all the hydrogen, and so some reincarnation of the stars, albeit as smaller stars occurs. Such stars contain amounts of elements other than hydrogen and helium. The sun is one such star. The heavier elements, sometimes called metals, can only have come from a previous supernova. The earliest stars were giant stars with a very short life but gradually the population of stars within a galaxy becomes dominated by the smaller G type stars.

Stellar regeneration can be seen in the stellar nurseries where stars are forming and igniting and are surrounded by a clear zone which is embedded in a moderately dense dust cloud. The question arises, if the new stars are from the primordial hydrogen helium mixture, why have they taken so long to form. The fact that they are surrounded by dust suggests that these are second, or later generation stars, formed out of some massive supernova that was the collapse of a giant megastar. It has been hypothesised that of the first stars formed, there was a predominance of giant stars. Such stars consume much of their hydrogen dowry very quickly before exploding as Type2 supernovas. Which then poses where does the hydrogen come from which is involved in the formation of these second-generation stars. Clearly some must come from the inefficiencies in hydrogen consumption before exploding as a supernova. But an interesting possibility is that the energy density at the heart of The supernova is sufficiently intense to force any nucleons there back into energy which then condenses back into newly minted nucleons. This reproduces in miniature the events of the Big Bang. Being formed in one reasonably closed space could account form why so many G stars are in binary formations.

It is becoming increasingly likely that many of the G stars have planets, as well as traces of atoms other than hydrogen and helium. This clearly indicates that they must have come from a prior supernova. If supernovas could regenerate pristine nucleons this would prolong the life of a galaxy before its light became extinguished.

This does also mean that galaxies with a lot of G stars will have a lot of dust. Possibly half the mass of the galaxy could be dust. This is one form of dark matter, which is not disputed. All this suggests strongly that our sun is a second-generation star. That it was born in a cloud of dust and its gravitational energy induced the nearby dust to form into planets. But more significantly, the solar system must be surrounded by a shell of very thin dust. This has relevance when considering the source of the background microwave radiation

Nevertheless galaxies must have a reasonably fixed life span before they cease to emit light. If the life of stars were fifteen billion years, then at the end of thirty, or sixty billion years at the longest, the galaxy would become virtually invisible. Many of its stars would have either exploded as supernova (The current estimate is that a supernova is likely in any galaxy at approximately one in every 100-200 years) Over a thirty billion-year period this could be almost 20% of its original collection of stars. Huge quantities of dust must be created with would obscure most of the light from its remaining smaller stars. Thus very distant old and dying galaxies would be invisible, but each one could still act as a gravitational lens. However the dust in these galaxies would have to be almost uniformly spread which given interstellar distances, is unlikely.

This also suggests that there must also exist dead galaxies composed mainly of the iron clinker of stars perhaps with a black hole at the centre but which are invisible to telescopes. However the occurrence of dead galaxies is only possible if the age of the universe is very substantial. The expansion of time hypothesis provides such a time span. But the universe would have to be old, at least thirty to forty billion years old, to contain such galaxies.

It could be argued that these ancient galaxies couldn't be as old as that as the life of the proton is only an estimated 30 billion years. But these dead or dying galaxies would be travelling at relativistic speeds, such that their internal time would be substantially slowed. Their energy output would lessen, and so relatively the available acceleration energy would be less, but they would be close to coasting along at very high speeds

One other startling effect of galaxies is their very efficiency in holding their stars together as a group whilst being accelerated to distant space by the expansion effect of the Hubble force, H_T. This is despite such occurrences as galaxy collisions, exploding supernovae in binary pairs of stars, or jet propulsion by energy emissions that result in jets of incandescent mass that can stretch for up to fifty thousand light years. There is, so far, no recorded case of an isolated star burning brightly yet far removed from any galaxy. Individual stars may deviate from the galaxial plane but the size of the deviation relative to the radius of the galaxy is trivial.

The stars start initially in the very large clouds of hydrogen and helium randomly mixed that is contracting to form the proto-galaxy. The clouds are at a very low temperature. Gravitational contraction draws the cloud together. The heat energy density rises warming the cloud. Bits of the cloud condense to form proto-stars stars of various sizes. Within the core of these proto-stars the density rises further and with it locally the gravitational force rises. Further gravitational contraction occurs. This slows time somewhat, causing loss of mass. The slowed time weakens the strong force that holds the sub nuclear particles together. (Force is inversely proportional to the square of the pace of time). But the nucleons are also losing mass from generalised time slowing. The system is therefore stable in its own time frame. This strong force, with gluons as its force particle, still has to oppose charge repulsion within the nucleus. The inference is that quark charge falls with time slowing. This must help to a degree with the fusion process. That is time slowing causes some of the mass to revert back to energy and the resultant energy output hastens the rise of the internal temperature. This is the basis of so called gravitational heating.

Stars with a high or very high helium content will warm up more slowly as their heat capacity is higher. They are less efficient at radiating away their load of energy. Their life is correspondingly shorter before they must explode as a type 1 supernova. This is the explanation of the finding that type 1 supernovas can vary from having no hydrogen visible in the spectrum to having a significant amount of hydrogen.

Initially in the proto-star there will be some gentle heating, the result of universal time slowing. With the shrinking in volume of the proto-star gravity will cause central compression and raise the core temperature. Eventually the temperature of the proto-star reaches levels such that the thermal velocity of the atoms at the centre of the proto-star is high enough for relativistic time slowing to occur. This causes the release of some of the atom's mass as energy. But eventually equilibrium is reached where the radiant energy opposes the gravitational contraction. Fusion is simply a compensation mechanism for sustaining the energy output without incurring an enormous energy debt.

For the larger mass stars with a higher gravitational strength the scenario is rather different. One requirement is that the radiant energy will have to be greater to oppose gravitational collapse into a black hole. But the initial key effect will be compounded by the gravitational effect on time. Gravity slows time. Slow time causes more mass to be released as energy, including heat energy. Because of the thermostatic effect controlling the temperature of a hydrogen-burning star, the increased energy output will cause more mass to move at relativistic velocities but the temperature will be held steady. The shedding of mass will also release large quantities of gravitons which will slow time further within the depths of the massive star. This will slow time further still and so cause the release of more energy from shed mass. The system is a vicious spiral that results in the exponential increase in fuel consumption. The core is consuming hydrogen at an enormous rate. This increased energy output penetrates the cooler outer coats of the larger stars so they appear hotter. As the spiral deepens so the rate of energy production increases. The star is accelerating the burning of its fuel at a higher and higher rate. The hot core enlarges. Very soon it will have consumed the bulk of its huge hydrogen store at its core, producing vast quantities of helium, perhaps as much as ten solar masses of helium. This huge amount of helium will contract but such a large mass of hot helium is very unstable and it explodes as a giant supernova, a Type 2 supernova.

Thus the symbiotic relationship between time causing mass loss together with gravity causing the slowing of time condemn the very large stars to a very short life, a life measured in a few million years compared with the many billions of years for the smaller G type stars. It is small wonder that such G stars are the commonest stars observed in our and our neighbouring galaxies.

For lesser sized stars, G stars like our sun, there comes a time when the inner core of the star starts to run short of hydrogen fuel. Convection currents prevent some of the lighter mass hydrogen outer coat from sinking into the core. As a result the internal radiant pressure is unable to oppose gravitational contraction and so the core contracts. This raises the temperature and eventually helium fusion occurs. The radiant pressure from this blows off the outer coat of cooler unburnt hydrogen, which expands to form a Red giant cloud (red because its temperature is lower). This results in the production of carbon, nitrogen and oxygen. But the dominant product is oxygen.

When the red cloud disperses what is left is a small very compact helium-burning star, the white dwarf. Carbon formation requires each atom of helium to lose 0.055% of its mass. In contrast each atom of helium needs to shed 0.068% of its mass to form oxygen. For nitrogen formation the situation is a little more complex since essentially it requires three helium atoms plus two protons or their equivalent to form an atom of nitrogen. There is no immediate explanation why the preferred end result of helium fusion is oxygen. There is a similarity with hydrogen fusion to form helium. In both cases it requires 4 units of the fusing material to form the end product. En route there will be formed subsidiary lesser mass material, deuterium and tritium for the hydrogen process, carbon and nitrogen for the helium fusion process. The 0.068% mass reduction for helium to form oxygen may be compared with the 0.726% of mass that the protons have to shed when forming helium. That is the increase in velocity to shed the mass to form the larger atoms is comparatively small once helium has been formed.. This means it requires only a relatively small rise in temperature to achieve this. This small margin contributes to the instability of helium burning stars and their proneness to go nova.

White dwarves are small compact very hot stars but with a mass that is only a fraction of its parent hydrogen burning star, which could be a mega star. It is fuelled by helium. Because it has a higher temperature than that of its parent the radiant pressure is correspondingly higher. This should cause an expansion of the star but in the process of formation the volume of the star has shrunk significantly. This means that the gravitational pressure locally is much higher. A similar effect would be seen on earth if the volume of the earth shrunk so that its radius was halved. At the surface of the earth the gravitational acceleration force would be quadrupled. The radiant pressure in the white dwarf has to combat the increased gravitational force, and the mutual tension between these forces is thereby increased. This system is fundamentally unstable and any break in the gravitational strength as by a localised violent heat release from a runaway fusion process would result in the catastrophic explosion. This is the Type 2 supernova.

Before then the core of the white dwarf would also be under intense radiant energy pressure compacting the centre, shortening the pathways between the fast moving helium and other atoms increasing the probability of fusion in its own right and forming progressively bigger and bigger atomic elements. Each fusion process releases more energy, warming up the process and accelerating it even more so that a vicious spiral then develops. The spiral ends when iron is formed as fusing iron consumes more energy than its releases. It is a net absorber of surplus energy. The core becomes progressively more and more laden with iron.

When the white dwarf explodes as a supernova the iron core is intensely compressed as the explosion rips around the core at close to the velocity of light. Pressure waves fluctuate throughout the whole star adding to the compression. If the star were big enough the delay in the explosion completely encircling the core would cause the core to be spat out like squeezing a pea out of a pod. Such an effect is seen in the Crab nebula where the neutron core is a pulsar that is offset from the centre of the nebulous cloud. This type 2 supernova exploded just under a thousand years ago in 1058 A.D.

Neutron stars and pulsars: A few years ago radio astronomers in Cambridge noticed very regular but very brief spikes of radio waves coming from space. Known affectionately as signals from LGM (Little green men) the source was soon identified. Soon other similar regular spikes were detected coming from other directions. They were quickly established that these were from tight beams of radio waves that appeared to describe large circles that would be substantially bigger than the earth's orbit. The radius of the circle described by the beam cannot be determined except that it is most likely to be bigger than the earth's orbit as, for some of them, they are to be seen throughout the year whilst the earth orbits the sun. It was soon established that the beams were coming from spinning neutron stars.

The expelled core from the supernova is very hot but progressively cools down and with no continuous fresh supply of radiant energy it contracts under gravity. The original compression due to the implosion of the iron core of its parent supernova would have driven the electrons into the protons neutralising the charge on the proton so that a dense compact mass of neutrons forms as a sort of giant atom, akin to a Bose condensate. (Atoms when cooled to a fraction above absolute zero condense behave as a single giant atom above which time is so slowed that light almost comes to a standstill). Free neutrons have a life expectancy, which is measured in minutes. They only have a long life when inside the nucleus of an atom or something similar. Incidentally this short shelf life is one of the reasons why in the very first stages of cosmology there was an excess of hydrogen formation over helium. Protons could exist independently but the neutrons had to find a partner or partners to form helium and had to find that partner quickly. If they missed the partnership they disintegrated into energy particles.

If the core of the white dwarf is small enough, less than a few solar masses the resulting outcome is a neutron star. The gravitational strength of a neutron star is insufficient to crush the neutrons into energy. Nevertheless the gravitational strength is extremely high, sufficient to slow time down significantly. This will cause the neutron star to lose some mass as energy. This is the source of the energy that fuels the pulsar radiation beam. Such a situation occurred with the supernova that generated the Crab nebula. This has a neutron star, a pulsar, which is substantially off centre from the expanding cloud of supernova remnants. The pulsar has a frequency of thirty hertz. The pulsar is rotating thirty times a second. The equator cannot move faster than the speed of light. This means that its equatorial circumference is at most $1/30^{th}$ of a light second, or ten thousand kilometres. For comparison earth has an equatorial circumference of a little over 20,000 kilometres. The radius of the Crab pulsar is less than three thousand kilometres, less than half that of the earth. This may be compared with the original size of its parent star, with a radius of at least three light seconds. In forming the neutron star the radius has been reduced to 0.3% of that of its original parent star. The volume has been reduced to 0.33% or 0.027% of its original volume, with much of the reduction due to compaction into neutrons.

The quantitative statistics of neutron stars are of some interest. Neutron stars cannot have a mass greater than four solar masses. Anything bigger and the gravitational pressure is sufficiently strong to crush the star into becoming a black hole. A solar mass black hole has a radius of about 3 km. A two solar mass neutron star one millionth as dense as a black hole would have a radius of ~600 km. Its circumference would Be 3.7×10^3 km or a little of 0.01 light seconds, a third of the circumference of the Crab's pulsar. If 0.01 seconds was the period of a signal coming from such a spinning neutron star then the equator of that star is travelling at the speed of light, and the pulsar frequency would be 100 Hz. If the radius was a little less, that is it was slightly more dense than one millionth of that of a black hole, neutrons at the circumference would be travelling a slightly less than the speed of light. But the implications of this high speed are of some significance.

Gravity slows time. At the surface of a neutron star the gravitational strength is very substantial. There, time is slowed significantly and this will cause the loss of some mass. The loss is energy, which would heat up the neutron star so opposing further gravitational compression. At the equator of a spinning neutron star the velocity of the surface neutrons is close to light speed. There this velocity would slow time further and so cause further loss of mass. Some of that energy would be gravitational energy, which would add significantly to the external time slowing and so further loss of mass at the equator. As the loss continues the radius would shrink but as angular momentum is preserved so the spin rate would increase. Meanwhile the equator is surrounded by a dense belt of gravitational energy, so great that it behaves like an event horizon of a black hole. Nothing can escape from the equator. On emerging at the equator the energetic photons as electromagnetic waves would be refracted towards to pole (The mechanism of the refraction is discussed in Chapter 6). If the refraction is too severe the photons will be bent into the body of the neutron star. But neutrons cannot absorb energy (otherwise their mass would increase). And so the photons are reflected back into the stream of photons going towards the poles.

As the spin rate increases so the equatorial belt of near light speed mass spreads towards the poles. The zone of time slowing spreads and consequentially mass is lost there. The resulting slimming increases the spin rate further. Eventually the neutron star becomes spindle shaped and then a tapering cylinder. All the energy is funnelled towards the tapering tips of the neutron star. But at the poles radiant energy can still escape.

The energy supply is prodigious. If the beam covers the earth's orbital area the energy density per square meter is such as to amount to equivalent of several million kilograms of mass per second. Such a rate of energy production can only come from conversion of mass to energy. By definition neutron stars cannot generate energy from atomic fusion. That is there must be a different mechanism. The time expansion effect on mass is the only currently available solution. That is the existence of pulsars is proof positive that time slowing causes mass to revert to energy.

The energy produced would cover the entire range, from gravitational energy to gamma radiation, depending upon the rate of mass conversion. But gravitational energy would be around 2-3% of the energy production. Neutron stars should therefore generate energy waves that cover the spectrum. The Crab Nebula contains a pulsar and also very weak Gamma rays and X-rays have been detected as coming from the Crab. Most of the nebula is very cool; too cool to be generating such energy so long after the supernova that formed the Crab in 1058 A.D. Logic therefore is that this radiation is coming from the pulsar. Light should also be produced and the source of the pulsar has been linked to a light source that is significantly off centre from the Crab nebula. Since by definition, the light cannot come from normal nuclear fusion, there has to be another way of generating electromagnetic radiation. This method has to be continuous to sustain the light.

However a spinning light source cannot describe a circle of light if the source is on the axis of rotation. There must be an additional factor or factors to account for the pulsation and the tightness of the light beam. In this context a tight light beam is one that has expanded by a minimal amount after a very long journey. If the light beam deviates by one thousandth of a degree from the axis of rotation then over a distance of five hundred light years the circle described would be 2.4 light days. This is five times as large as the entire solar system.

The gravitational field around fast spinning neutron stars must vary, being strongest at the equator because of the gravitational energy released by the atrophying neutrons. The field would resemble a bi-lobed swelling that completely encircles the star At the pole there would be a very narrow funnel of minimum gravitational pull. That is energy would travel down the funnel of the gravitational vortex. The lobes of gravitational energy extend outwards for some distance from the star and this would act rather like a Fresnel lens constricting the beam, like that of the lens of a searchlight, as a very strong gravitational field bends the light. The gravitational field would be spinning with the neutron star's rotation. The photons, light or radio, would be refracted from side to side as they passed up the vortex and so emerge as a very narrow beam. Inevitably there would be a very small angular offset. That is the time mass relationship with its energy production of gravity, and electromagnetic radiation obviates the need for some form of very large external magnetic field close to the pole of the pulsar to account for the narrowness of the beam.

Black holes. Do they exist? Black holes were originally considered as occurring where the density of mass was such that the gravitational force arising from this mass would be such that it would require a velocity greater than the velocity of light for anything to escape, including photons. The limit of this field is the event horizon. Anything inside the event horizon is trapped and cannot escape. This includes energy particles. The general equation describing the event horizon (also known as the Schwartzchild radius) of a black hole is

(Equation 7,1) $$r = 2GM/c^2$$

where G is the gravitational constant and equals 6.672×10^{-11}, M is the mass in kg, and c is the velocity of light in metres. A million solar mass black hole should therefore contain about 0.1% of the galaxy's mass But all is not as it seems. A ten fold increase in the mass of a black hole results in a ten fold increase in radius, and also a thousand fold increase in volume and a hundred fold reduction in density. According to this equation a million solar mass black hole should have a relative density that is a thousand billionth of that of a one solar mass black hole. But this assumes that mass remains unchanged. Clearly this cannot be true.

An alternative explanation therefore is necessary At the event horizon the intensity of the gravitational pull is the same irrespective of the mass of the black hole. Its energy level is therefore unchanged. Following a four-fold increase in mass, the radius would have doubled (maintaining the ratio M/r^2 which is an intrinsic part of the gravity equation). The surface area would have quadrupled, increasing the amount of gravitational energy radiated out into space. The volume enclosed by the Schwartzchild radius has increased by a factor of eight. The density would have halved. The surface area would have increased fourfold. But gravity is a normally a simple function of the density. To maintain the same level of gravitational energy at the horizon per square metre of the horizon's surface the total gravitational energy must have also increased by a factor of eight, although the mass has only increased four fold. The only source of this excess gravitational energy must be from the breakdown of some of the protons, crushed out of existence. This in turn is a positive feedback situation. The excess gravitational pressure crushes more protons until all the protons are converted to energy. The excess gravitational force would also cause time to stop.

One can calculate what the radius would be if the earth was compressed to the density of mass that makes a black hole. Earth would have a radius of 8mm. The sun would have a radius of 1.5 km., whilst a million solar mass black hole would be 10^6 times greater and so would have a radius of five light seconds. (The sun has a radius of approximately 2.3 light seconds). A hundred million solar masses would have a radius of five hundred light seconds. This is the same at the earth's distance from the sun. So far no one has identified a black hole with more than a few million solar masses.

If the event horizon is outside the black hole mass there are a number of ambiguities. According to relativity theory, time, that is the second, is dilated to infinity at the event horizon. Effectively therefore there can be no velocity within the black hole. There is total stasis and all forces cease to act. The temperature should be at absolute zero since that requires motion. There can be no rotation as momentum energy is so attenuated by time dilation that the body comes to a standstill. Without rotation there can be no luminous jet but these jets have been observed. The acceleration force should also cease, which would mean that the black hole cannot accompany the rest of the galaxy as the galaxy moves outwards as part of the expansion of the visible universe. Gravitons cannot radiate out into space, as that requires motion from within the event horizon. Furthermore if gravity ceases to act it cannot dilate time. It follows that the event horizon cannot exist outside the mass or its residue that makes the black hole.

If the putative event horizon is deep within the body of the black hole then the density of mass is insufficient for the body to be a black hole. It can therefore radiate all forms of energy including gravitational energy

At best this positions the event horizon at the surface of the black hole. Even then there are problems as time within the black hole would be infinite and there would be no gravitational force. It follows that there must be some mechanism, which prevents gravity contracting the mass to the required density of a classically described black hole. .

Another related problem is how a black hole can achieve a gravitational effect of a million plus solar masses. There must be some very powerful gravitational force arising from the centre of the spiral galaxies for these galaxies to have that shape. In this it is akin to the solar system. Galaxies start as a globular mass of stars, up to a hundred billion of them, but gravity is the only force which can move the stars to form a quasi planetary system rotating around a central dominant body. When a black hole starts it may have the mass equal to ten to thirty solar masses. This would be the case if it arose from a hundred Solar mass star that had developed into a Type 2 supernova. For a small black hole to a distant star the gravitational pull would therefore be less than that of the black hole's parent star, although locally around the mass the gravitational force would be immense. It follows that to build up the gravitational strength to the equivalent of a million solar mass a putative black hole must have a gravitational force that is excessive relative to its mass.

Following from the principle that time expansion causes a reduction of mass and the release of gravitational and expansionist energy, time slowing around the putative black hole could account for this, even if pure gravitational pressure did not crush all the nucleons. With the implosion of the iron core of the type 2 supernova its density and so gravitational force locally would be extremely strong, though the initial heat of the core that came from the supernova would oppose this. As it cooled so the contraction effect of gravity would increase adding to the local gravitational force. The relativity equations show that gravity slows, that is it expands time. Time expansion causes mass to reduce, the resulting energy release being heat, the acceleration force and more gravitational energy. The release of energy would generate radiation pressure, which would slow up the gravitational contraction, just as the radiant energy within the sun prevents the sun from contracting down to a black hole. A dynamic balance would be achieved. But some of that radiant energy would escape causing cooling and so further gravitational contraction. The rate of cooling would be determined by the temperature of the mass and the slowing or expansion of time in and around the putative black hole. The excessive gravitational force would cause the putative black hole to pull in nearby stars. The resultant collisions would increase the mass and so its total gravitational energy output, so attracting more and more stars and so on until a huge mass would exist.

A galaxy consists of up to several billions of stars, of various sizes. The rate of supernova formation has been estimated as one per two hundred years per galaxy. The majority of these would be Type 1 supernovas but a significant minority would be type 2 supernovas, the progenitors of putative black holes. Over a billion years a galaxy should have experienced up to a hundred thousand Type 2 supernova, each with its putative black hole remnant. Their increased gravitational pull would make them more responsive to the gravitational pull of another black hole. If the latter had achieved a certain size the newly formed putative black holes would be quickly drawn to the central black. Its angular momentum from circling around the centre of the galaxy would be conserved, so that it would spiral at increasing velocity towards to the dominant mass. As it neared that mass it would enter into the slowed time zone, and its relative velocity would decrease, that is relative to the rest of the galaxy, but the velocity would be the same, relative to its own slow time frame. As it neared the dominant mass so the gravitational force between them would be increased. For the incoming putative black hole time would be slowed even further causing the release of even more gravitational energy and reduction of its mass. Eventually there would be a collision. But it would be a very gentle collision relative to our time, due to the excessive time slowing around each of the black holes. Whether this would produce gravitational waves of sufficient strength to be felt well outside the volume of the galaxy is doubtful. But the dominant black hole would have even greater gravitational energy. This should have happened many times in the older, that is the more distant, galaxies.

In a number of galaxies a jet of light is to be seen arising from near the centre of the galaxy and extending almost half the length of the galaxy. Occasionally there may be a smaller jet arising from near the same source and pointing in exactly the opposite direction. The jets are narrow, not much wider than the stars they are passing but slowly fanning out as they approach light extinction. One remarkable facet is that the jet is as bright as the stars it is passing as well as being extremely straight.

It is possible to construct a model to gain insight as to the quantitative implications of the energy used by these jets. Assuming it extends for less than half the radius of a typical galaxy—this would make the jets 20,000 light years long. It starts as a point and ends as a slim base, say one light hour radius (One light hour is two thirds of the distance between the sun and Saturn) after travelling for so long. I.e., the collimation is very effective The surface area of this light cone would be approximately 7 x

10^{15} square light seconds. One square light second represents 9 x 10^{16} square metres. That is 1 joule per square metre equals the equivalent of 1 kg of mass expressed as energy per square light second. For comparison the sun, radius 2.3 light seconds, has a surface area of 66 square light seconds. The energy output of the sun results in the solar constant of 1400J/sec per sq. metre at a distance of 500 light seconds from the sun. This corresponds to ~10^{27} billion Joules per square light second at the solar surface. This is the energy equivalent of ~10 billion kg of mass per second. If the jet from the black hole has the same brightness as the sun it is emitting energy as radiation the same as the sun (although the mechanisms of producing the energy may be different). The cone of light with a surface area of 7.x 10^{15} square light seconds is therefore the result of transforming 7 x 10^{25} kg of mass per second as energy. This corresponds to ~3 x 10^{32} kg of mass per year or approximately 150 solar masses a year. (A solar mass is 2 x 10^{30} kg.) These masses have been totally converted to energy and that the process has been going on for at least 20,000 years for the light cone to be so long. That is the light cone represents at a minimum the energy that would be derived from over two hundred thousand solar masses. But the light beam may also contain mass, which is emitting this radiation. The shape of the cone suggests there is a very special lens working to prevent to dispersion of light just as occurs with radio waves from pulsars. That is the immense gravitational field around the black hole is also acting like a Fresnel lens, just as it does for neutron stars that form as pulsars. Although this model is very crude it does indicate that the light beam seen perhaps a billion light years away represents a massive amount of energy. There is only one body, which can produce such mass, and that is the gradual deflation and extinction of a large black hole.

A black hole therefore is simply a spinning ball of energy, that is held together by the immense gravitational field. In this respect it represents a recapitulation of the original ball of energy that powered the Big Bang. Energy turned into mass has reverted back to energy.

One can therefore construct a life history of a putative black hole. It starts as the impacted core of a Type 2 supernova. This blob is ejected at high speed from the supernova. It has to be; otherwise the detonation of the supernova would have to be instantaneous around the core over a very large surface area. But the speed of light would inhibit the simultaneous detonation The blob is predominately compacted iron vapour. It is very hot and the radiant energy initially opposes any gravitational contraction, just as it does in our sun. But the heat is being radiated away and is not being replaced. The opposition to gravitational contraction lessons and so it shrinks. That raises the temperature somewhat and the contraction lessens until that heat is also radiated away. And so the cycle is repeated. The intense gravitational field slows time and this causes some of the mass to revert to energy, including gravitational energy. That is the gravitational energy is greater than the mass would normally produce. When this is high enough it opposes the radiation of other forms of energy. The entrapped residual radiation energy eventually blocks further gravitational contraction, but only when the object is close to forming, if it has not already formed a classical black hole where nothing can escape. The density is such that the event horizon is either at or very close to the surface. The result is a small body with a grossly excessive gravitational pull that is capable of attracting distant stars and other putative black holes. This it does and so the body becomes increasingly massive. But gravitational energy is being radiated away, albeit very slowly. But as it attracts additional mass so the event horizon enlarges (proportionately to the square of the additional mass) so the rate of radiation of gravitational energy increases.

Internally with the loss in some gravitational energy the pace of time starts slowly, very slowly to increase. The increase in the pace of time induces some of the emaciated nucleons to increase their mass by absorbing some of the trapped radiant energy, and also some of the excess gravitational energy. The increase in the pace of time also increases the magnitude of the force exerted by the trapped radiant energy. Unless the object can absorb more mass and regain its gravitational strength it becomes more and more unstable as the internal pressure builds up, although some of that radiant energy will also escape with the gravitational energy that is being radiated away. Then the black hole will rupture at some point with a massive emission of energy and some mass. As the internal pressure lessons gravity will regain its dominant hold and the emission will cease. This is the typical gamma burster. A variation is if the rupture is big enough gravity will not quite regain its full dominance and the result is a continuous deflation producing an immense beam of light mixed with some mass. In this it is imitating the pulsar but on a much grander scale. Because the rate of release is much greater than that of the pulsar of a neutron star system the radiation will be of shorter wavelength, that is a beam of light. Overall a comparative analogy is that it will be like the slow deflation of a football from a small puncture with the puncture site exhibiting a jet of pressure.

But the end result is the same. Unless the putative black hole can be continuously fed by attracting more mass from the rest of the galaxy it faces gradual dissolution. The event horizon is either at or just below the surface so that with the radiation of gravitational energy some acceleration energy will be released. The acceleration energy is very weakened by the slowing of time induced by gravity. But the mass inside the black hole will be reduced as much of it has been transformed into compacted weakened energy so that it requires less push from the acceleration energy. The result is that the black hole is able to keep up with the motion of the rest of the galaxy.

One striking phenomenon that arises over the interaction between time and mass and that is that Newton's laws of gravity, or rather the equations that derive from them are not valid if there is significant change in the rate of time. A time slowed mass will have a greater gravitational pull than a non-time slowed mass. It is this extra pulling power than is responsible for the type 1a supernovas, when the white dwarf pulls hydrogen from its twin partner (in Binary systems) which eventually triggers the detonation. Without this extra pulling power the white dwarf which has a mass that is less than its original parent star, would not be able to attract the hydrogen from its partner when its parent was not able to do so.

There is one other oddity about the effects of the absence of time within a black hole. If there are no stasis energy particles, photons etc could move about but in the absence of time they could be in two places at the same moment. Alternatively the energy particles could all be merged into one giant condensate, something like the

Bose-Einstein condensates that occur when hydrogen is liquefied to just above absolute zero (where times also virtually ceases).

Quasars: or quasi-stellar objects are star like objects but with a brightness that far exceeds that possible for a normal or even a very large star. They are usually found in or close to the centre of a galaxy. Their energy output is massive. Their origin and fate is described in the previous chapter.

Jupiter: The planet Jupiter has a curious weather pattern. Across the equator are large storms that tend to form cellular patterns. The storms create winds of several hundreds of kilometres per hour. Yet its distance from the sun precludes solar energy as being the source of such violent behaviour. The question then arises could the energy released by any time slowing be the fuel source for this weather. The answer is surprisingly positive.

Jupiter is a big planet, with a diameter of 142,700 km. Its rotation period is ten hours. This means that the velocity on the surface of Jupiter at the equator is 12.5 km per second or 4.15×10^{-5} of c, the velocity of light. From the standard relativity equation this velocity would expand time by 8 parts per ten billion. Such an expansion would reduce mass by 8 kg per ten billion kg of mass. But each kilogram of mass is worth 9×10^{16} Joules. That is per second 7.7 million Joules of energy will be produced per kilogram of gas that at the equator is being accelerated to Jovian speeds. One standard barometric earth pressure (15 lbs./sq. inch) corresponds to the pressure generated by almost eleven tons per square metre. At one thousandth of an atmosphere pressure would correspond to a mass of 11 kg per square metre. That is the outer atmosphere of Jupiter down to a pressure of one thousandth of earth's atmospheric pressure, a distance of a few kilometres, would contain enough mass to generate 80 million Joules per square metre per second. This figure may be compared with the earth's solar constant of 1400 Joules per square metre per second. The 80 million figures is based on the assumption that all the gas at the equator has started moving from rest. But even if one hundredth of one per cent of the gas started effectively from rest, that is drawn from higher latitudes, this still equates to 8000 Joules per second per square metre.

Clearly some of the high-speed gas will spill over to more favourable latitudes where the rotational velocity is less. Their molecules will try to regain their lost mass by absorbing energy. In doing so there is created a significant energy gradient between the equator and the latitude in question. The stage is therefore set for the cellular pattern of weather disturbance and very powerful vortex movements within the atmosphere. If the vortex by chance becomes larger enough the velocity around the rim of the vortex could be high enough that time slowing occurs there. This would result in sufficient energy production to maintain the vortex indefinitely. Could this be the explanation of Jupiter's famous red spot?

All though this description is a gross simplification of the problem it demonstrates that the weather pattern on Jupiter could be accounted for by its sheer velocity of rotation and the effect that has on the mass time relationship. Not all the energy produced will be thermal energy. Some will be gravitational, some will be accelerational, some will be electromagnetic including radio wave energy, some could be infrared, and some could even be light.

The Moon and the Earth. The moon consists of rock that has the same composition as the earth's oldest rocks. Its mass is 1.5% of that of the earth. It clearly came from earth. Its relative size in anomalous compared with the moons of other planets. It clearly came from the earth and was the result of large meteor hitting the earth. This meteor was around 200 kilometres across and its impact velocity was huge.

Prior to impact the earth was largely covered by water as there were no deep oceans. The water depth was up to two kilometres. The impact created a giant tsunami a mile or more high that raced around the earth. The meteor crashed through the earth's crust, through the upper viscid layers of the mantle to reach the deep layers where rock was melted at a temperature of over 1000°C. Molten rock was thrown up at such a velocity that much of it was thrown into space where it eventually cooled to form the moon. The mass ejected was over 2% of the earth's mass although not all reached escape velocity. The tsunamis raced around in opposite directions, met, recoiled and raced back to the crater reaching it some many hours later. At the crater the waters faced a pit bigger than the Pacific Ocean and over 100 miles deep. The floor of the crater was exposed molten rock. The water cascaded into the pit with a force that took it several miles below the surface of the molten rock. There the water flashed into high pressure steam which propelled more molten rock towards the nascent moon. Meanwhile the crater had created a huge updraft of hot air which in turn generated a gigantic tornado. More water poured into the crater together with the slow moving molten rock. The oceans of the world poured into the crater and were boiled dry. The steam carried by the updraft reached the stratosphere where it spread and cooled condensing back to water. Then followed a huge deluge, as the entire oceans crashed down, washing out any rock in the atmosphere back to earth. On reaching the ground the water flowed back to the crater to be boiled again. This continued until enough magma had filled the crater and the flow of heat eased.

With the loss of so much magma under the earth's crust the crust collapsed breaking in to plates. Land levels fell by over two kilometres. On the far side of the earth the giant super continents, Luarasia and Gondwanaland were fractured, the fragments were pulled slowly toward the crater by the underlying flowing magma creating the continents that we now see. They had not travelled far before most of the crater has filled in. That slowed their movement more but they are still drifting towards the crater site even today.

The meteor strike was when the earth was about 1 billion years old. At that time the heat produced from the decay of Uranium 235 was fifty times faster than it is today (Uranium 235's half life is such that it has passed through almost six half lives since the earth first formed). There would therefore have been enough heat energy to melt the rock almost to the present depth of the earth's crust.

The moon enables a quantitative test in principle of the effect of time in releasing gravitational energy. The expansion of time causes mass to be loss at a rate of 2.3×10^{-18} of its mass per second. The total energy output is divided into acceleration energy, gravitational energy, and almost certainly some heat. The ratio of their constants between the expansion constant and G the gravitational constant shows that gravity uses 12.5% of its available energy after any possible heat losses. Gravitational energy is radiated out in all directions into space and simple geometry shows that the earth intercepts 6.85×10^{-5} of this radiation. With this background it is possible to create a mathematical model to show the effect of the expansion of time and consequent energy production in predicting the change of height of the ocean (say the mid Pacific ocean,) when the moon is directly overhead. Table 6,1 shows the steps of the model.

The gravitational energy arrives as a beam with a cross sectional area equal to the cross sectional area of the earth. Its gravitational energy content per square metre can then be calculated. The gravitational energy passing through a strip of area 1 metre wide and of length equal to the diameter of the beam in metres would fall on a similar one metre wide strip over half the circumference of the earth when the moon is directly overhead.

Every twelve and half-hours (approximately) the moon creates two tides. One on either side of the earth, for balance so stopping the earth precessing, so that at any one site the tidal interval is around 12.5 hours. Because of the rotation of the earth and its spherical shape the effect of the beam would be that of a shallow sinusoidal arch, of length that is half the circumference of the earth with the apex of the arch gently moving from east to west. Simple arithmetic enables the height of the arch to be calculated in energy terms. From then on it is a matter of comparing that energy into the equivalent of that needed to produce by a column of water with a base of 1 square metre. This resulted in a prediction of the height of the tides at mid ocean as being around 5 metres but almost certainly much less because of the viscous drag of pulling water from areas of low tide. The model assumes a circular orbit of the moon around the earth, and it also excludes any effect from the sun's gravitational energy –i.e. it excludes neap tides. It also excludes the effect of the slight variation in distance of the moon due to the earth's curvature. The model assumes that the water absorbs all the gravitational energy. But water is a poor absorber of gravitational energy as witness the different tidal heights seen at inland stretches of water, whether the Mediterranean Sea or the Great Lakes of North America.

Despite all these reservation the model turned out to be surprisingly robust. If $1/3^{rd}$ of the energy (mechanical efficient 33%) released by time expansion was shared just between acceleration energy and gravity and there was no frictional or viscosity loss the change in mid ocean height with the tides would be 7.3 metres. For comparison the tidal height at the Galapagos Islands is 2 metres.

Furthermore despite the simplicity of the model the concordance between the effect of predicting what should be the result if time expansion is taking place and what happens is striking. This is particularly so when considering that the starting point of the model is the mass of the moon of over 7×10^{22} kg

Comets and Asteroids: The distinction between an asteroid and a comet is not a hard and fast one. Asteroids are generally thought of being made of rocky material that is usually held together rather strongly, much like a lump of rock. They can be of any size but most commonly less than a couple of hundred kilometres across. Comets are usually much more fragile, a mixture of frozen chunks of water ice, Carbon dioxide ice, ammonia and sometimes low mass hydrocarbons. The nucleus often has more solid material such as free carbon and even some silicates. Some may well have a small solid rocky core. A comet striking the earth is likely to vapourise and explode in the atmosphere unless the comet was very large. The damage can be very substantial. In 1908 a comet of mass about 0.1 tons, and diameter less than 200 metres struck the earth at Tunguska in Siberia. It exploded high in the atmosphere and totally destroyed everything within a radius of 40 km, that is an area larger than Greater London or the city of New York.

Another comet of interest is Comet Shoemaker-Levy 9 that broke up and crashed into Jupiter in 1994. This comet was orbiting the planet Jupiter. It was more fragile than the Tunguska comet and had broken into fragments during a previous orbit when it passed an estimated 25,000-km from the top of the Jupiter's clouds. What is of interest is why it broke into so many fragments and generated a cloud of dust and many very small fragments in addition to the major fragments. There were 19 major fragments plus lots of little bits and dust. The fragments continued to orbit Jupiter forming a long train that took from the first arrival five days for all the train to get past Jupiter. The orbit was highly elliptical and had a maximum length of 50 million km (about one third of the earth's distance from the sun). From being near stationary at the furthest point of its orbit the comet was accelerated by Jupiter's gravity to a peak velocity of 60 metres per sec. Such a velocity would slow time sufficiently to cause the loss of mass the equivalent of 4500 Joules per kilo of the comet. Given the speed of change much of this energy would be in the form of heat, which in the middle of the comet would be trapped by the poor heat conduction of the comet's icy walls. Inevitably there would be some melting which would weaken the walls. The comet's trajectory around Jupiter would impose a mechanical strain on the outer wall furthest away from Jupiter leading to cracking and eventual rupture. The steam would act as a jet causing the fragments to separate. The tail end fragments would be slowed so opening a progressive gap. On the next orbital circuit the gravitationally induced acceleration would widen the gap so extending the train of fragments such that there would be as much as a 5 day interval between the first fragment and the last fragment crashing into Jupiter. The steam induce rupture of the comet would cause a certain amount of fragmentation of that wall forming a cloud of debris which would accompany the fragments (and which was seen).

During their final approach to Jupiter a tail was to be seen from each of the fragments. The tails were at right angles to the direction of movement, and emerging from the side of each fragment with the weakest walls. The distance was too far for the sun to cause this. (The thermal energy from the sun at this distance is insufficient to generate significant tails from other comets when they are at this distance nor does it generate clouds or tails from the icy moons that circulate around Jupiter.) Furthermore no tails have been reported as being visible from this comet when the fragments were leaving Jupiter during the first half of their final orbit. That is the energy required to produce these tails was from more heat and so more steam, being generated due to time slowing and mass energy conversion by the renewed burst of acceleration as the comet or rather its fragments approached Jupiter again.

The alternative suggestion is of tidal forces causing the break up. But this would not generate a cloud of debris. It also implies a differential in the distribution of the energy along the surface of the comet. But the comet was too small for this to be significant. Thus comet Shoemaker-Levy offers an example of the effect of time slowing on mass.

What happened to Shoemaker Levy must happen to all comets as they accelerate at high speed towards the sun. This would also explain why recent photographic close ups of comets show that some of their tails erupt from the rear of the comet, the portion not exposed to the sun. Increased velocity and time slowing with mass loss generates enough heat for low temperature and low pressure steam to form the cometary tails.

Technical Note

A model to predict the effect of the expansion of time on the Moon as it affects the earth. The model is based on three assumptions.

(a) That 2/3rds of the energy released by time slowing ends as heat., that is the mechanical efficiency for the production of useful work is 33%.,

(b) 12.5% of the remaining energy is used as gravity, the rest is acceleration energy (ratio 7:1).,

(c) That half the energy arriving on earth to raise the tides is spent in frictional losses dragging water from areas of low tide to the high tide area

Table 7.1. A simple algorithm to show the maximum possible effect of the moon's gravity of the height of the tide in mid ocean when the moon is directly overhead. It assumes that large oceans absorb all the gravitational energy that is arriving to produce the tides. The table assumes that the sea water has a density of 1.1 kg/litre. Half the energy is subtracted due to frictional losses. The calculation is then the sum of a simple arithmetic series of the energy required to raise successive layers of a square metre of water that is 1 mm thick in building a column of water that equals the energy supplied by the moon.

Item E., Assumes that 4% of energy released by time slowing is gravitational energy

Item J. The energy input when the moon is directly overhead

Item L The receiving area per tide is 12.5/24 x the earth's circumference owing to the moon's movement during the tide

Item M. The volume under an arch is 0.5 x 3.14 x base area x height. The area corresponds to energy supplied for the whole tidal period, the base is the receiving area, and the height is therefore the peak energy and corresponds to the energy of high tide

Item T. The column was calculated as a series of layers each 1 mm thick by 1 sq metre. The energy of each layer was therefore its mass times earth's gravitational energy times the distance. This becomes a simple arithmetic series, the sum of which gives the peak high tide.

Note: If the gravitational fraction from time slowing is 6.25% the peak high tide becomes 6.45 m

	Identity	Action	Result	Units
A	Mass of Moon		7.35×10^{32}	kg
B	Time slowing energy	$A \times 2.3 \times 10^{-18}$	169050	kg/sec
C	Energy in Joules	$B \times 9 \times 10^{16}$	1.52×10^{22}	J/sec
D	Energy per 25 Hours	$C \times 25 \times 3600$	1.37×10^{27}	J/25 hr
E	Gravity's fraction (4%)	$D \times 0.04$	5.47×10^{25}	J/25 hr
F	Earth Moon Distance		384400	km
G	Shell area	$F^2 \times 4 \times 3.14$	1.86×10^{12}	km^2
H	Energy per sq Km	E / G	2.95×10^{13}	J/km^2/25 hr
I	Earth's Diameter		12756	km
J	Energy supply in strip Earth's diameter x 1 km	$H \times I$	3.76×10^{17}	J/strip/25 hr
K	Energy input per tide	$J/4$	9.41×10^{16}	J/strip/tide
L	Earth's receiving area	$I \times 3.14 \times 12.5 / 24$	20861	km^2
M	Peak energy of arch	$K \times 2 / (3.14 \times L)$	2.87×10^{12}	J/km^2
N	Peak energy per metre^2 after 50% frictional loss	$0.5 \times M / 10^6$	1.44×10^6	J/m^2
O	Density of sea water		1.1	kg/litre
P	Earth's gravity		9.8	m/sec/sec
Q	Volume in layer 1 mm x 1 square meter		1	litre
R	Mass of layer of water	$O \times Q$	1.1	kg
S	Potential energy of layer (Height is 1 mm)	$R \times P \times 1$	10.78	J/mm
T	Numbers of layers (Sum of arithmetic series)	Solve $n(n+1)/2 = N / S$	516	
U	High tide if 33% mechanical efficiency		5.16	Metres
	High tide if 50% mechanical efficiency		6.45	Metres

Chapter 8

Mythologies ancient and modern

Ancient myths, dark matter, the background microwave radiation,
stretching of waves, the hot spots, dark energy, modern cosmology

Every civilisation has developed myths of the origin of the cosmos. With one notable exception the myths involve astronomical objects, the commonest being the sun and moon. The exception is Norse mythology whereby creation arose out of icy chasms and the first gods were the frost giants. This myth is possibly a long folk memory of the last of the Northern Ice Age. The less sophisticated the society the simpler are the creation myths. The Doghon tribe in Mali believed that there was a god who created the sun from a pot of clay. The Australian aborigines have a vague concept of a dreamtime, whilst the Polynesians believed that the earth was created on the back of a giant turtle. The ancient Egyptians had a polytheistic cosmology. The Sky god was Nut who gave birth to Osiris, Isis and other gods. They believed that the Milky Way was the goddess Nut. Her mouth was the constellation Gemini.

Her genitalia was where the Milky Way appears to divide slightly, by the bright star Deneb. In the evening of every spring solstice Nut swallows Ra the sun god—the Gemini appear to overcome the Sun in the evening of the spring solstice. Nine months later at the winter solstice Deneb appears over the horizon just before the sun rises. Nut has given birth to Ra.

The Sumerians had their chief god Anum, god of the sky and Enlil god of the earth. Their sun god was Shamash whilst Ishtar, his sister was goddess of storms. The Babylonians chief god was Apsu who through Tiamet begot a number of gods and goddesses. When Apsu threatened to kill them all there was a rebellion. Apsu and Taimet were killed by their grandson Marduk. He cut up Tiamet's body and used it to make the stars, the heavens and the oceans. Similar bloodthirsty creation myths occur in other societies.

In India there was a primordial man, Purusha, who was killed by his offspring and body cut up so that his eyes produced the sun whilst his soul produced the moon. In India there are other creation myths. Brahma arose from within an egg. When he emerged from the egg he split the shell. He then made the sky out of one half of the eggshell and the earth out of the other. Thereafter follows a perpetual cycle, every 4320 million years whilst Brahma is awake the earth passes through a succession of ages which decay into corruption. Then Brahma sleeps for another 4320 million years during which the universe dissolves into a watery chaos. When Brahma awakes he recreates the universe and the cycle is repeated. There is no beginning and no end. This myth strangely echoes a modern proposition of an endless cycle of the universe expanding from a singularity as a Big Bang and then under gravity contracting back into a singularity as the Big Crunch.

In China there were the emperors of the Northern and Southern Seas who met in the land of Chaos ruled by the emperor Hundun. To thank Hundun for his hospitality they offered to give him the seven holes in his head that humans have. When they drilled the seventh and final hole Hundrun died but from this seventh hole came the entire universe.

In MesoAmerica there was an elaborate calendar of four suns each sequentially lasting about 500 years. There were the creation gods Tonacatecuhtli and Tonecaciihuatl who had four sons, Tlatlauhqui, Yayauhqui, Quetzalcoetl and Huitzilopochi. At the end of each sun's period the earth was destroyed. After the fourth sun Quetzalcoetl and Huitzilopochi were ordered to create a new world as well as a sun and moon. Quetzalcoetl was the god of the calendar that had supreme importance in the lives of the Mesoamericans. In their cosmology the gods had to be appeased and this meant building large pyramids on which human sacrifice took place.

The Greeks had their pantheon which was related to the stars. The Milky Way was the semen spilt when Kronos (the Roman Saturn) castrated his father Uranus with an adamantine sickle. His children, the leader of whom was Zeus, destroyed Kronos in turn. The Roman cosmology largely followed that of the Greeks. It does however contain a startling description of a possible fate of the Earth. Phaethon, son of the Sun God Phoebus, persuaded his father to let him drive the chariot carrying the sun across the sky. In his book Myths and Legends from Ancient Rome Kenneth McLeish writes "but now, as the horses began their downward plunge, he suddenly saw the starry plains ahead were filled with monsters; the Scorpion, Bootes, iron clad oxen, the Crab with its pincers poised. Fear loosened his fingers and he dropped the reins"

The sun was burning and destroying the earth and evaporating the seas. Eventually Jupiter (Zeus) has to kill Phaethon with a thunderbolt. This is an uncanny forecast of what one day will happen when the sun becomes a red giant which will destroy the earth and evaporate the sea.

Michael Jordan in his book Myths of the World describes another 50 more different creation myths from around the world. The vast majority invoked a key role for astronomical objects in some shape or form.

Polytheism gradually gave way to monotheism. In the Middle East arose a religion based on a single god, Yahwah. This was the ancient Jewish religion and from it arose first Christianity and then Islam. Their creation myths are to be found in the book Genesis.

Characteristic of all these pantheons was the enforcement of the beliefs through armies of warrior priests. This occurred in Ancient Egypt, and in Greece. Anybody challenging the beliefs was killed or had to die. Socrates the greatest of all Greek philosophers was forced to commit suicide after challenging the existing beliefs. But such blood thirsty enforcement was to persist, through the Crusades, the bloody spreading of Islam across North Africa, the various European religious wars, the Inquisition and so on.

Gradually however the cosmology changed to one of great celestial crystalline spheres increasingly perfect as one moved outwards. Heavenly music, the music of the spheres permeated the sky. Even Kepler, one of the founding fathers of modern astronomy tried hard to link his calculation of the planetary movements to music.

But Earth was still central. The stars and the sun rotated over the earth under the influence of God who could stop the movement if necessary. Thus in the Old Testament book of Joshua it is claimed that the Sun and the Moon were stopped whilst the Jews killed their opponents from another tribe. Even if this is interpreted as stopping the rotation of the earth the consequences are rather drastic. At the equator when viewed from space the Pacific Ocean is moving at approximately 1000 km/hr due to the earth's rotation. With more than a billion cubic metres and each cubic metre of water weighing a ton, a sudden arrest of the earth's rotation would have created a tidal wave, which would have flattened the Rocky Mountains. It all seems a trifle disproportionate simply to resolve the battle between two small warring tribes on the other side of the world.

Slowly with the discoveries by Copernicus, Galileo, Tycho Brahe and Newton a heliocentric universe controlled through gravity became impossible to deny. The great heavenly spheres had to go. Nowadays we recognise that the sun is only one of a very large number of similar stars in our galaxy. In turn our galaxy is just one of many billions of other galaxies.

But the denial of myth still creates enormous antagonism as Charles Darwin discovered with his newfangled theory of evolution. And in more modern times those who opposed Lamarckism in Communist Russia.

Myths are created to provide simplistic explanation to be believed by the gullible when there is difficulty in otherwise explaining observed phenomena. When pronounced by people with authority they become dogma. Any challenge therefore questions the authority of the great and the good and is not to be tolerated. Modern astronomers have not been slow to create their own myths. Propositions, initially put forward tentatively, become the accepted order of things particularly if they seem reasonable in the light of the available knowledge at the time. A typical example is the concept of the Hubble constant described in an earlier chapter. But there are other myths.

Some Modern myths

Dark matter: There is confusion over this term. It initially was by the astronomer Zwicky to account for non-luminous matter that he believed permeated through a galaxy or a cluster of galaxies and which, he proposed, generated sufficiently high gravitational energy to account for the anomalous velocity of the outer galaxies. Then, within galaxies, it was applied to dust, hydrogen clouds, brown stars, failed stars, that is masses which were too small to develop into full blown suns, as well as burnt out helium stars too small to form neutron stars.

Stars orbit around the core of galaxies. For a stable orbit of any body orbiting a gravitational mass the standard equation is

(Equation 8.1) $g = GMm/r^2 = mv^2/r$

where g is the gravitational force between the two masses M and m (both in kilograms) that are distance r metres apart, v is then the velocity of the orbiting body in metres per second. Applied to the earth's velocity around the sun the equation gives the orbital velocity of 29.4 km/sec. This can be confirmed by the following calculation. The earth's orbit takes a year and the earth-sun distance is known. Assuming the earth's orbit is completely circular (it is only slightly elliptical) the circumference becomes known and the resulting calculation yields a velocity of 29.8 km/

sec. The difference is presumably attributable to the varying size of r due to the elliptical nature of the earth's orbit.

The equation applies to any orbiting body. It follows that when there are several orbiting bodies each at different distances from the centre, the ratio of distances should match the inverse ratio squared of the velocities of any pair of stars orbiting a galaxy. When applied to the average velocities of stars orbiting in at different distances from the centre the equation held up reasonably well. There was a short fall of 50%. This was attributed to these cold non-luminous masses. It was then held that this dark matter accounted for half the mass of a galaxy. With this allowance it was equations like this which enabled astronomers to give an estimate of the mass of a galaxy treating individual stars as having a mass equal to our own sun. It therefore came as a compete surprise that the outermost stars, where the gravitational force would be correspondingly very weak, had velocities that were over eight times that predicted by the equation. The immediate inference was that the gravitational pull was much greater than expected. That is there must be a large amount of another kind of dark matter within a galaxy which is contributing to the excess pull and so causing the remote orbiting stars to require much higher velocities in order to stay in stable orbits. The excess mass has been estimated as being some eight time the previous estimate of the mass of a galaxy.

There is a problem though if there is a large amount of non-luminous matter congregated at the centre then the velocity of the stars at middle distance should also be excessive compared with the stars close to the inner core. That has not been reported, in fact it was the opposite, which provided the data enabling the mass of any galaxy to be calculated. If the dark matter were congregated to the more peripheral parts of the galaxy its quantity, eight times the total previously calculated mass of the galaxy, would be so substantial as to cause significant dimming of the stars from the central parts of the galaxy. If it was non-homogeneously distributed throughout the galaxy then the galaxy's spiral or elliptical shape could not be maintained. Nevertheless there seemed to be no other explanation for the excess velocities.

One approach has been to calculate the actual increase in g required producing these anomalous velocities. It was shown that g had to increase by a factor equal to between 10^{-9} and 10^{-10} m/sec/sec. It was thus proposed that Newton's Law, which predicts the exact strength of g with distance, deviates from normal when the g force is very low, that is at very long distances. This is the MOND hypothesis. The association between distance and gravitational strength is a long and cherished one that has never been proven to fail before. This explanation has not gained general acceptance. In the absence of any other explanation the dark matter or rather the dark mass hypothesis gained general credence.

There is though a force, magnitude 4.86×10^{-10} m/s/s, which is doing work, pushing galaxies along. This is the acceleration force H_T. It is logical to presume that a force, which propels galaxies, should be able to propel stars. When the gravitational force is strong the effect of the accelerating force will not be noticeable and be lost in the inaccuracies of measuring distance when looking at distant galaxies. It is only when g is very weak that the accelerating force becomes prominent

It is suggested that the force H_T exerts a stabilising effect on the orbit of any body particularly if there is a fluctuating g force because the orbit is not completely circular. The origin of this force from within the nuclei of atoms that make up the orbiting mass places the force in perfect position to exert its effect. It follows that there is no need to adjust Newton's laws of the behaviour of gravity but merely add the Hubble acceleration constant.

A recent expansion of this dark matter theory has been based on observations from the Hubble Telescope. These were that in an angular area four times the size of the moon there are apparently a large number of gravitational lens in the very distant universe in areas where clusters of galaxies seem to be colliding. The more distant they are the more granular is the appearance of the distribution. Closer to our Galaxy the arrangements seemed to be clumpier. Closer still means a number of billions of light years away. This has been interpreted as showing the distribution of dark matter. There are profound difficulties with this interpretation. The first is that the dark master seems to be concentrated towards the edge of the universe, which is contrary to the large-scale uniform distribution of galaxies, the isotropy, throughout the visible universe. It was this isotropy that was the underlying driving force for postulating the Big Bang theory. Furthermore if gravitational lens are due to masses of dark matter it is a puzzle that why are none reported within the nearest billion light years of our galaxy. Yet dark matter has been invoked to account for the anomalous velocities of the outer galaxies of nearby clusters of galaxies. What actual has been seen is lots of pairs or triplets of galaxies with the same red shift. The inference then is that there is a single source whose light has been refracted by a gravitational source giving multiple images. But the numbers do not add up. The alternative explanation is that of refraction by large clouds of gas of thin density that have failed to form into stars and galaxies. Thus there is no need to invoke a peculiar distribution of a peculiar substance called dark matter.

If this is true then the dark matter hypothesis is fatally flawed and should be confined to history like all the other cosmological mythologies conjured up by the sages of classical history.

The background microwave radiation: In 1965 two engineers, Penzias and Wilson, working for Bell Laboratories in New Jersey were involved in the building of a microwave receiver for microwave communication with orbiting satellites. They kept getting a background hiss despite all their efforts at cleaning their giant receiver. Eventually they consulted two physicists Peebles and Dicke in Princeton University. These two physicists were deeply interested in a proposal by three other physicists Gamow Alther and Bethe that the universe started with the Big Bang and that there should be some radiation equating to a temperature a couple of degrees or so above absolute zero. Since the observed microwave noise corresponded to radiation at about this temperature it seemed natural to attribute this to the Big Bang

We are constantly being bathed in very low level of microwave radiation. The source of this radiation is somewhat enigmatic. It comes to us from all directions and apart from the hot spots is at the same frequency, corresponding to a very low temperature. It is this non-direction that creates the problem in the identification of its source. It has been described as the after glow of the Big Bang. But what is glowing? The afterglow of any radiation requires that the energy was absorbed by something and is subsequently released after the primary source has discontinued supplying energy. For example the afterglow of a jet engine is from the release of heat from the hot metal casing of the engine and the heated air molecules cooling down when the engine is switched off.

The calculations of the amount of residual radiation were based on the energy density that would result if space had a limited volume and there was some form of reflection to cause uniform mixing of the energy, or that space itself would have absorbed some of that energy. At the time although it was realised that the universe is big, just how big was an estimate. The estimated size turned out to be grossly too small. For the energy to come from all directions means that there must be some form of reflecting surface But the energy density and so temperature would depend upon knowing the proper size. The need for a reflecting surface for the microwaves imposes a limit upon the size of the universe. A key observation is that the strength of the signal is the same in all directions.

But the space enclosed by the observable universe is expanding at a rate that the edges of that space are moving at the speed of light. That is when the energy of the Big Bang was released it radiated outwards to the edge of space and would have travelled with the edge of that space. The radiant energy would have left the middle zones of space completely unless there was something able to absorb some of that energy. The analogy is a light beam from a searchlight. When the source of that beam is switched off there are no photons left near the source, although far in the distance those photons are still travelling at the speed of light. It follows that with the edge of space moving so fast there has not been time for any reflection of that original release of energy.

One suggestion that space is that space itself is capable of absorbing energy. This requires some controlling mechanism for the subsequent release of that energy. But the release is constrained by the volume of space and its rate of change or expansion. In order to have a uniform temperature across the whole universe at any one time each part of space must "know" exactly by how much the local space has expanded at any moment of time. This is in order to release the appropriate number of photons. The rate of release determines the energy and temperature of the emitted radiation. Once released photons can summate with other photons, but since they are in wave form they must be in phase with each other. Given the size of the universe and the limitation of the speed of light this is impossible.

The frequency of the radiation also defines the temperature. So the constancy of the background microwave radiation requires some form of control strategy so that photons can be emitted at a later date. But the problem does not stop there. When the universe was half its age its volume was less and so the energy density was higher. This would have resulted in a higher temperature with photons to match. Those photons would then have to be absorbed and subsequently released at a lower temperature as space expanded. This would have been a continuous process since the Big Bang. This process is fundamental if the background microwave radiation is a remnant of the Big Bang. The hypothesis is testable.

The hypothesis is that space absorbs heat photons and subsequently releases them when the source has gone. The released radiation is at a lower frequency. The sun releases a prodigious number of heat photons. The sun is also moving around of the centre of the galaxy. It follows that the space where the sun was, say one thousand years ago, should have absorbed an excessive number of heat photons compared with more distant parts of the heavens. That space should now be releasing some of those heat photons and this should be clearly visible as a hot spot trail delineating the path of the sun over the last one thousand years. No such trail is evident in the maps that show the hot spots. If space were able to absorb microwave photons we would not be able to feel the full heat of the sun.

Another problem is that there should be progressively hotter microwaves coming in more distant space, but which have taken their time to arrive because of the limitation of the velocity of light. But the background microwave radiation does not show this.

The emitted huge amounts of radiant energy were at very high frequency. Much of this energy condensed to form first quarks and electrons, then nucleons and finally atoms. (Under the current Big Bang theory the singularity was too small for it to have contained all the electrons and nucleons that make up the mass that is the universe) Although some of that radiation from the Big Bang could have been low frequency waves if the background microwave radiation is a remnant of the unused energy the cut off is astonishingly clean. One would have expected a tailing off to higher wavelengths

The remote possibility is that the singularity that sparked the Big Bang is still emitting large quantities of low frequency radiation that is spewing into the universe. But this does not account for the non-direction of the microwaves.

A major problem with the Big Bang origin of the microwaves is the uniformity in intensity of the signal no matter which direction one looks and whether one looks during the day or during the night. Reflection is not possible, because of the size of the universe. Adsorption of microwaves by space with subsequent emission at a lower frequency universe is excluded for lack of evidence.

With these two exclusions there then should be a difference in intensity depending upon whether one is looking towards any putative site of the Big Bang or looking towards the edge of the universe. There should be some kind of shadow effect as the earth absorbs the microwave. But none has been seen.

One final point the intensity of the microwave signal is the same in all directions. East, west, north, south, the signal strength is the same. If this is a relic of the Big Bang this must mean that the photons responsible have all travelled the same distance, otherwise the inverse square law would come into play. This positions the earth as being at the centre of the universe. Observational data shows that this is clearly not so, although the centre cannot be defined.

There remains one other difficulty. The Temperature of the microwave radiation is ~2.7K corresponding to an energy density of 10^{-14} J per cubic metre. Halving the radius reduces the volume by a factor of eight, that is the temperature would increase eight fold for each halving of the radius. If one takes the radius of the universe as 13.7 billion light years (almost certainly a gross under estimate) it is possible to calculate the radius of the universe when the temperature was 10^8 degrees. Above this temperature protons cannot exist. The radius would have been ~50 million light years. This may be compared with the radius of our Galaxy of ~50,000 light years. The hot photons would have fled leaving an empty massless void of radius fifty million light years at the centre of the universe. No such void has been seen.

One is left therefore with the implication that the microwaves are originating from all around us but not from extra-galactic space, as there is nothing there to contain or absorb and subsequently release the microwaves. It cannot come from the molecules that make up the earth and its atmosphere since microwave radiation has been observed by satellite probes looking away from earth. In any case the temperature is too cold.

The microwaves therefore must come from within our galaxy since that is what surrounds us. Our galaxy contains a lot of dust that is at the temperature of around 2.7K. The conclusion must be that it is the dust within our galaxy that is radiating the microwaves as reflecting the temperature that the dust particles have. Given the large surface area of the dust particles relative to their individual masses clearly they would cool to absolute zero unless that heat energy was replenished. But that dust is experiencing time slowing. That is it is shedding mass as energy. The dust particles will have their own gravitational energy component, and their own acceleration component. That is it is emitting energy to fuel both of these. Logic suggests that in doing so it is generating its own heat component.

The release of electromagnetic energy leads to an interesting speculation that as the mass of the dust particles slowly decreases so does the charge on the individual atomic nuclei. But charge is quantum dependant, that is the energies being released are also quantum dependant. This includes gravity. Could this be the long sought lead in to a quantum explanation of gravity? The orbits of the electrons of the atoms that make the dust would then be pulled tighter towards the nuclei of each of the dust atoms. But according to quantum theory they can only do so in jumps. A change in orbit will cause the emission of electromagnetic waves of specific frequency related to the change of orbit. The continuing loss of mass due to the continuous time slowing means that energy so released replaces the energy lost in radiation and so keeps the temperature of the dust particles at a constant level.

What is more germane is that a background microwave radiation of some sort is an obvious and predictable consequence of the effect of time slowing on mass if there is a lot of dust in our galaxy. What is clear is that the theory that the background microwaves are remnants of the energy of the Big Bang should now be discarded.

The dust that surrounds us is a remnant of the giant supernova which spawned the solar system. Inevitably it would be minor irregularities in its distribution. Such irregularities are to be seen in the images of stellar nurseries. Some of the dust would form clumps. The emitted radiation would therefore be slightly warmer than other nearby regions. And this is exactly what is seen.

One is therefore forced to the conclusion that the observed background microwave is arising close to the solar system. Its continued presence is further confirmation of the hypothesis that time is very slowly expanding.

The stretching proposal. To resolve the dilemma that the observed microwave background has a very narrow width of frequencies, corresponding to a black body radiation at a very specific level it has been proposed that this is due to space expanding or stretching the microwaves. Frequency is the velocity of light divided by the wavelength. It follows that an increase in wavelength must be matched by a decrease in frequency. But each wave contains the same Planck unit of energy, and the wave is travelling at the velocity of light.

Radiation waves cannot extend their wavelength without violating the principle of an absolute constant velocity of light in space (relative to the time frame) as was discussed in an earlier chapter. Neither can waves merge to create a single wave of longer wavelength when two waves are travelling in the same direction and are following each other. That would require one of the waves either to accelerate to catch up with the preceding one or the preceding one to slow down to allow a catch up.

Another argument against stretching is that light from the most distant supernova has the same colour spectrum as light from a nearby one. There is no stretching of the colour light waves from the supernova during their very long transit across space. There is the red shift, that is some stretching but that applies only during the actual emission of that light caused by the velocity of the emitting body as the photon leaves that body. More importantly that body, if it is moving, must show a red shift in all its emanations. That is if the red shift is due to stretching during the light's transit there is nothing to show that the emitting supernova, or indeed any distant quasar is moving. But they should have a red shift if they are moving, as they must be if the universe is expanding. This concept of the red shift being due to stretching whilst in transit contradicts the concept of an expanding universe. But if there were no expansion then there would be no stretching. Attributing the red shift to the stretching hypothesis thus creates a paradox.

Brightness of any object depends upon the frequency of photon emission and the amplitude of their waves and the waves themselves being in phase so that they can summate. Stretching would therefore reduce the brightness of the light from any distant object. But this would cause another violation of normal physics. The inverse square law would be inapplicable if during the passage of light through space the space itself was stretched as well as the light waves. The inverse square law relationship between brightness and distance would not hold. What applies to light waves must also apply to gravitational waves. That is Newton's Laws of gravity would be violated. Gravity would be "dimmed".

The stretching hypothesis also negates the idea of dark mater. If there was stretching, the red shift from stars on opposite sides of a rotating galaxy should be the same. But standard observations show this is not so. Indeed the differences in mean velocities on either size of a galaxy that have been observed have been used to calculate the Hubble constant. And of course it was these differences in velocities, identified by their red shifts that provided the only justification for claiming that there is dark matter, and lots of it.

All in all there is no evidence for the stretching hypothesis. It was an interesting hypothesis that on further examination is found to create too many difficulties including breaching several fundamental laws of physics. It should therefore be relegated to the same treatment as the epicycles of yesteryear.

The hot spots. More recent measurements from satellite probes have disclosed that there is large number of areas where the microwave radiation indicates a fractionally higher temperature than the surrounding areas. These have been attributed to remnants of the original radiation from the Big Bang. There are a number of problems about this explanation. Radiation cannot stay still by itself. As its name indicates it radiates the energy away at the speed of light. That is it would have mixed and spread throughout the universe. Radiation from even hotter zones will mix but will not oppose the dissemination of the lower wavelength radiation. The analogy is that of the radio broadcasts spread across everywhere and two sets of radio waves at different frequencies do not interfere with each other. Any continuing emission from any site indicates that there is something still emitting the radiation. In space the rate of spread is not determined by the temperature difference. All radiation spreads at the same speed, c, the velocity of light. Any sustained difference means that there is a continuous supply of energy. So what is glowing?

One problem in identifying the sources of these hot spot emissions is that their distances are not known. More recent observations have shown that many of them have an angular size equal to or greater than the angular size of the moon. This severely limits their distance.

If they were as far distant as our nearest neighbouring group of galaxies the angular size predicts that the size of the emitting zone would be considerably greater than all the galaxies in that cluster. Anything that big that is glowing must have mass and the mass would be immense. Its gravitational effects would be obvious.

The moon's diameter is 0.9% of its distance. Therefore anything of the same angular size its diameter would be 0.9% of its distance. Even if the hot spot zones were just on the edge of our Galaxy at 100 light years distance at a minimum means that the diameter of the hot spots would be almost one light year. This may be compared with the diameter of the sun, which is a little over four light seconds in diameter. That is the zones would be 130 million times bigger than the sun. This is possible given that observations of distant galaxies show large dusty coronas. This does not explain their continued ability to emit radiation, as they should cool down to the temperature of absolute space. A possibility is that the dust clouds in the corona are reflecting back the heat from the galaxy but this does not explain the uniformity of temperature.

A more likely explanation is that just as the microwave radiation is coming from the dust clouds within our galaxy so the hot spots are also from the dust clouds. Their slightly raised temperature would result if the dust clouds were beginning to coalesce under the influence of their own gravity and what is being seen is the beginnings of gravitational heating as the radiation emitting dust particles gather together and so warm up.

What seem to have been ignored are that all the planets, planetissimals, comets, Oort Clouds etc. arose from a vast dust cloud that enveloped our sun and came from a past supernova. Whilst the innermost parts of the cloud under the sun's gravitational field condensed into a disc and then formed the planets, the outer most parts of the dust cloud would be exposed to a very much weaker gravitational pull from the sun. They would stay in cloud form with the possibility that the innermost part of this outer zone formed the Oort clouds. If they were twice the distance to Pluto, at ten thousand million kilometres their angular sizes would indicate that the diameters of the hot spots are only about five light minutes. This would be big enough for feeble gravitational condensation to occur. This concept is hypothetical but is also a consequence of time slowing and its effect on mass.

Thus the idea emerges that beyond the Oort Clouds are thin but large patches of dust surrounding our Solar system. This thought is not so strange as it first seems. Telescopic observations of so called stellar nurseries shown stars being "born" inside huge dust clouds. Those dust clouds must have arisen from previous supernovas. There is no reason why our sun should be different. Such thin dust clouds could also explain the small variations in peak magnitude in the light from very distant type 1a supernovas when the supernovas are travelling at the same speed. It should be recalled that the orderly expansion of the universe requires that type 1a supernovas receding at the same velocity should be the same distance and so should have the same peak magnitude. The only exceptions to this rule are where the host galaxy of the supernova is being pulled or deflected by some local Big Attractor, or is on a collision course with another galaxy, or that some of the supernova light is being intercepted whilst en route.

Whatever is the cause of the hot spots their angular size precludes them being outside our galaxy. Just as finding the Holy Grail was mythological wishful thinking so it is just mythological wishful thinking to claim that the hot spots are the residua of the Big Bang. The reality is more prosaic.

Dark energy. Galaxies are receding at velocities, which increase with distance. This is the Hubble expansion. It takes a lot of energy to accelerate a billion solar mass galaxy. The energy has been called dark energy. Its precise source is not known but is alleged to arise from within space. Under the Big Bang theory all galaxies are more or less the same age. Since the more distant galaxies are moving more quickly they must have received more energy. But the Big Bang hypothesis is very precise that during the inflation period energy was equally distributed through the volume that existed during this inflation period, as energy was travelling faster than light. The result of this homogenisation of energy is that the distribution of galaxies is uniform Homogenisation of energy contradicts the concept that the more distant galaxies received more energy, or that distant space releases more energy than nearby space. The Hubble acceleration constant H_T also has it that all galaxies are receding at an acceleration of 4.86×10^{-10} metres per second. The distant galaxies clearly then have been accelerating for a much longer period of time.

A parallel situation arises when considering stars near the rim of galaxies. They too are accelerating at a rate of slightly less than 10^{-9} metres per second. This has been attributed to dark matter rather than dark energy. But the coincidence of figures is striking.

More recently the two Pioneer space probes have been found to be accelerating at slightly less than 10^{-9} metres per second. The acceleration was unexpected and is unaccounted for. There is no apparent fuel source to provide the energy for the acceleration

There is an old saying once is happenstance, twice is coincidence, and thrice is circumstance. Something is making all three different sizes of mass all accelerate, and accelerate at the same rate, but they are in different positions in the Universe and are unconnected. These masses are the receding galaxies, the stars near the rim of galaxies and the Pioneer probes, and their sizes are monstrously different. The obvious solution is that mass generates two forms of force. Gravitational force and acceleration force. That being so dark energy arising from space then disappears. The energy is arising from within the mass just as gravitational energy arises from within the mass. And that energy is being continuously supplied. It must be in order to maintain the acceleration.

There is however a more obvious objection to extrinsic dark energy accelerating a massive structure such as a galaxy. An external force capable of moving a million solar mass black hole would strip the galaxy of its stars and totally destroy its architecture. Examples of this may be seen in the break-up of galaxy structure when that galaxy is exposed to a strong external force such as the gravitational pull from another galaxy when the two galaxies are on a collision course. The disruption occurs when the galaxies get too close to each other. They could be many light-years apart but the gravitational forces render the structure chaotic. All this has been observed.

Dark energy must therefore be considered an interesting speculation unsupported by fact, and the phenomena that it was trying to explain have an explanation that has a base well rooted in observation. The energy for the expansion of the universe is coming from the mass that is being moved. That energy is the result of time slowing on mass. Dark energy must therefore be classified as mythology much like the dark forces alleged to belong to necromancers.

The Big Bang theory: In its present form the Big Bang theory must be wrong yet in a number of details the theory gives an explanation which is compatible with normal science. Although the Big Bang theory is examined more deeply in Chapter 10 there are several key points to show that it must be a myth if only because of its frequent invoking of what can only be called magic. Magic may be defined as using mechanisms that do not obey any known law of physics to produce an effect. In the story the conversion of Cinderella's pumpkin into a stagecoach is by magic

The first objection is that the Big Bang theory uses magic to explain the uniform distribution of galaxies. The inflation hypothesis whereby photons travelled within a limited volume at velocities far in excess of the known speed of light is an example of breaking the known laws of physics to produce an effect.

The second key objection is that the theory ignores the effect of gravity. The total energy, or mass equivalent of that original singularity was that of many billions of black holes. The strength of the gravitational field was such that no photon could achieve enough speed to escape from gravity. Yet gravity is one of the prime forces of nature. It must have existed along with the other fundamental forces of nature within that hypothetical microdot that the theory posits to have existed. That is the laws of physics were being violated again to produce an effect. This is invoking magic.

The third violation of known law of physics involves time. There simply has not been enough time for clouds of gas to condense, form galaxies, and for stars to go through their life cycle before exploding as supernovas and then for that light to take in excess of ten billion years to reach us. According to the laws of physics for a gigantic cloud of hydrogen to condense and eventually form a galaxy would take several billion years if the only available force were gravity. Then the stars would have to fire up and go through their life cycles. For all this to happen within less than three billion years must mean that there had to be a mechanism to accelerate the process. No such process is known within the laws of physics. That it the process must have involved some form of magic!

Although the Big Bang theory generated widespread acceptance at the time especially when it was coupled with the discovery of the background microwaves it has outlived its usefulness. It certainly concentrated minds wonderfully on the genesis of the universe and is very successful in explaining the current proportions of hydrogen and helium in the universe. Nevertheless the concept that the universe started as a singularity is based purely on an extrapolation backward in time on the current rate of expansion. The extrapolation was only stopped because it reached Planck length, the lower limit of length. There is no reason in principle why the extrapolation could not have stopped earlier, say as a result of radiant pressure or for any other reason.

The Big Bang theory is based on the assumption that the period of the second has remained unchanged since the beginning of time. If one allowed for the expansion of time then clearly what Big Bang theory would consider as very brief moments of time could have taken billions of years. Light would not have needed to exceed its maximum velocity. Nevertheless not all the problems disappear A reconfiguration of the Big Bang is given in chapter ten.

Chapter 9

Some Philosophical Musings

Seti, the oxygen mystery, a question of religion.

SETI: The search for extra terrestrial intelligence has been the cause of much waste of effort, money and resources. This has arisen because a simple but fundamental question was not asked. This is "What would be the power requirement to broadcast a radio signal so that it is heard throughout the year by someone on earth when the transmitter is fifty light years away?"

There are two ways of looking at this question. The first is an indiscriminate broadcast, much like the sun, radiating the signal in all directions. This is extremely wasteful. This is something that would be appreciated by any intelligent extra terrestrial organisation. But assuming that directional radio had not been worked out for any reason, it can be calculated what the signal emission to reception energy ratio would be. The solar constant is 1400 Joules of energy per square metre per second at 500 light seconds from the sun. That is the sun is emitting radiant energy at a rate of 65 million joules per second per square metre of its surface. At a distance of 50 light years, from the inverse square law the amount of energy delivered from the sun would be 1.4×10^{-10} J per square metre per second, a reduction by a factor of more than 4×10^{-17}. This is the ratio of energy from transmitter to receiver over a distance of 50 light years. The question then becomes what is the minimum energy per square metre capable of being detected on earth. The background microwave energy amounts to $\sim 10^{-14}$ Joules per second per square meter. At a generous estimate if unequivocal signals can be detected at energy levels as low as $\sim 10^{-18}$ J per square metre. The transmitter energy output required is that of about 10% of our sun. No physical material could tolerate such a rate of energy transmission without vapourising immediately.

The second way of looking at this is to assume that the Extra Terrestrials were able to use a directional radio broadcast. It still would have to cover the earth's orbital area around the sun if the signal is to be heard throughout the year. It also requires phenomenal accuracy at directing the signal aiming it specifically at planet Earth. That also raises other interesting speculations as to how do they know we are here. But the earth's orbital area is approximately one million square light seconds. Energy at one Joule per square metre of the earth's orbital area equals a mass equivalent of 1 kg per square light second. That is to cover the earth's orbital area, if reception was 1 joule per square metre per second, this would require the transmitter to send the energy equivalent of one million kilograms of mass per second. If the detector had a sensitivity to detect energy levels at 10^{-18} J/sq. m/sec the transmitter would need to send 10^{12} kg of energy per second, nine million watts. But what must also be taken into account is the mechanical efficiency of getting that amount of energy from the electrical source to radio waves. This would have to be at a specific frequency. There would then be consequent energy losses. This would be at least 90%. That is the power demand of the transmitter would be in multi-Mega watts. But more importantly this amount of energy transfer is not possible with any metal for a transmitter of any reasonable size.

But is their intelligent life anywhere in the universe? With something like 10^{30} stars, many with planets, statistically it is highly probable. Even if only one star per galaxy had developed intelligent life, this still leaves many millions of intelligent life sources in the universe. This is because there are so many galaxies. The sad thing is that they are all too far away for any contact to be made, or even for any useful detectable signal that they are there.

Nevertheless it is possible to have some idea of the requirements for intelligent life. It must be carbon based, since no other element has the flexibility to form the many various complicated chemicals that make up a living organism. For intelligence it requires speed of movement if only to develop strategies to avoid predators or to become a successful predator. Predation is essential to drive the development of intelligence. The fuel requirements for speed are oxygen plus some carbon based combustible material. The chemical reactions need water. Oxygen could not exist on such a planet initially as it is too reactive with other elements, such as hydrogen (for water), silicon (for sand), aluminium (for bauxite), or carbon (carbon dioxide) as well as the various metal oxides. The planet with life must have been seeded from a previous supernova with a large dowry of organic but inert chemicals to provide fuel for the initial stages of evolution. Examples of this dowry are the abundant hydrocarbon lakes seen on Titan. The oxygen when it comes must be diluted with another inert or reasonably inert gas –we have nitrogen-- as too much oxygen is profoundly toxic—it promotes the speed of chemical reactions.

In addition to rapid motility for significantly intelligent life to develop there needs to be developed tools and so the means to hold the tools. Whether the life form would be oviparous or placental is perhaps irrelevant. But some form of sex development would be essential for the rapid evolution of intelligence. If the life could develop wings as well as something to grip tools this would be a profound evolutionary advantage. (It is our misfortune that the development of the two-girdle system soon after the Cambrian explosion prevented mankind from developing wings as well as useful hands). Clearly the planet must have a stable temperature, not too hot to boil the water, nor too cold for water to solidify as ice. Periodic renewing of the surface by rainfall is another necessary prerequisite. But the main puzzle is the oxygen.

The oxygen mystery

"It is an old maxim of mine that when you have eliminated the impossible, whatever remains, however improbable, must be the truth." Sherlock Holmes in "The Sign of four" by Sir Arthur Conan Doyle

The origin of atmospheric oxygen and its concentration is a mystery. Our present atmosphere consists of a fraction under 21% oxygen; the balance is mainly nitrogen. Over every square metre of the earth there is two and a quarter tons of oxygen (Atmospheric pressure due to the mass of air above is 15lbs/sq inch, or 24,000 lbs. per square metre, 21% of which is oxygen). The surface area of the earth is ~10^{21} square metres so that there is a total of 2.25 x 10^{21} tons of oxygen as gas. Oxygen is the commonest element of the earth, accounting for nearly half (49%) of the total mass of the earth. Yet when the planet first formed all the available oxygen was used to make substances like water, sand and the many oxides, silicates and carbonates that make up the earth's non-gaseous mass. Geologically the oldest iron deposits that erupted late have the least amount of oxygen as oxides, that is all the available oxygen had been used up. Despite this very primitive life forms evolved. Their energy must have been provided by the anaerobic breakdown of the abundant hydrocarbons endowed by the parent supernova. To anaerobic organisms oxygen is a deadly poison. Yet at the time of the Cambrian explosion, when complex multi-cellular organisms suddenly appeared across the globe, atmospheric oxygen had to be at least half its present concentration and probably more. That is the atmosphere contained at least 10^{21} tons of oxygen. But then and only then could multi-cellular and multi-layered organisms evolve with cellular differentiation into limbs and sense organs etc. This they did with extreme rapidity. Too rapidly! The narrowness of the geological band carrying the earliest multi-cellular fossils compared with the extensive band carrying unicellular organisms, or sheets of single cell organisms points to a sudden change. These early multi-cellular fossils show the widest variety of fundamental body plans ever seen on earth. This wide diversity of structure is typical of evolution. When a substantial ecological niche opens up, life evolves rapidly to find the forms that can take the greatest advantage of that new niche.

It has been suggested, for want of a better explanation, that the oxygen arose as a result of the emergence of photosynthesis. But there are problems with this hypothesis. Photosynthesis relies on the breakdown of carbon dioxide, where the carbon is stored in a form that can be used as an energy source in the immediate future. Every gram of oxygen produced in this way leaves just over a quarter of a gram of carbon that is not joined with oxygen. This means that for every square metre of Pre-Cambrian rock there should be between 600 and 1250 lbs. of un-oxygenated raw carbon (between a quarter and half a ton per square metre.). But this has not been found in the oldest rocks, that is rocks that have not undergone subduction back into the earth's mantle layer. Furthermore photosynthesis has been continuing ever more vigorously since pre-Cambrian days but there is no evidence at all of a continuing rising oxygen levels since that time (approximately 600 million years ago).

Cells use up oxygen to metabolise their stored energy supplies. That is the excess oxygen to supply the atmosphere can only be balanced by the residual carbon of their carcasses when those organisms die. Furthermore that carbon has to be in non oxygenated form (lignin, oil, methane). Most of that residual carbon carcass deposit is in the form of carbonates, such as calcium carbonate or chalk. Effectively this means that there is less oxygen available for the atmosphere. There simply are not enough of these non-oxygenated materials to account for the 2.5×10^{20} tons of oxygen that is currently in the atmosphere. Such large deposits of unoxygenated or poorly oxygenated carbon that have been found are coals, tars, oil or methane deposits but none of these are found in pre-Cambrian rock. The coal in any case arose from plants that existed well after the Cambrian explosion and so is not a legacy of the oxygen generation.

Another problem then arises. If all the oxygen arose from photosynthesis than there must have been the equivalent of 21% carbon dioxide available. From its water-gas partition coefficient of 100:1 the atmosphere then would have contained at least 0.2% Carbon dioxide. The partition coefficient is the ratio of the carbon dioxide dissolved in water to the amount that is in an equal volume of air above the water. The ratio is that there will be 100 times more gas dissolved in the water than in the air above the water. This 0.2% figure relies on their being equal volumes of atmosphere and water. But the volume of water on earth is considerably less than the volume of air, that is with less water to dissolve the carbon dioxide the concentration in the atmosphere would have be higher than 0.2%. But even taking the 0.2% this is five times higher than the current concentration 0.04%. If this was the original concentration of carbon dioxide in the atmosphere it should have triggered a greenhouse warming effect. The gas is less soluble the higher the temperature of the water, that is the greenhouse effect itself would have triggered a massive outpouring of carbon dioxide previously dissolved in the water. This would have worsened the greenhouse effect and the end result would have been a very hot Earth similar to that of Venus. Life as we know it would not, could not exist. But it did.

An alternative hypothesis is that the oxygen came from photosynthetic breakdown of water. This would liberate hydrogen as well as oxygen. Cells cannot store hydrogen. If it is taken into solution and ionised in some way the resultant acidity would kill the cell. Ergonomically there is no benefit from metabolising water. To gain energy benefit the cell would have to be able to store hydrogen. Then when it requires energy it must join that hydrogen with oxygen. It is a chicken and egg situation as neither would be in the atmosphere should such a system start. In any case where has all the hydrogen gone? It would float to the surface of the atmosphere but then gravity would keep it there. To claim otherwise, that the hydrogen would escape from gravity would be to deny that star formation and indeed galaxy formation occurred. Both rely on the gradual gravitational condensation of vast hydrogen clouds. But little to no hydrogen has been found in the upper atmosphere. Furthermore if the oxygen came from water the atmosphere would consist of almost 40% hydrogen, but hydrogen and oxygen is an explosive mixture. Any lightening strike would take out not only the hydrogen but also the oxygen.

There is a theoretical possibility that the oxygen arose from deep with the earth's core. The core, which makes up to 40% of the earth's mass, is predominately iron. Its temperature is 6000+ degrees. It would take more than 10^{32} Joules of energy to achieve this temperature. Iron, molecular weight 55.847, has 26 protons the rest of the nucleus are neutrons. Astronomically it is rarer than silicon. Why the iron should be concentrated in the earth's core is a mystery unless it was formed there, Silicon, atomic weight 28.08 has 14 protons, the rest being neutrons. After oxygen it is the next most common element on earth. It exists as either silicates or sand, Silicon dioxide. Stripping the oxygen from the silicon and converting one proton per atom of silicon to a neutron, by adding an electron to the nucleus, and then forcing two atoms of the modified silicon atoms to fuse, would create iron. More significantly it would release approximately a ton of oxygen per ton of formed iron. To strip the oxygen from the silicon would take a lot of energy. More energy would be required to force an electron into the atomic nucleus. Yet more energy would be needed to generate enough compacting pressure to force the fusion. But the energy released from mass by time slowing would be ample. The oxygen released would very slowly force its way to the surface, carrying some of the newly minted iron with it. As it approached surface it would cool to below iron's boiling point (3000 degrees). Simple iron oxides would form, consuming some of the oxygen. At the end of the Pre-Cambrian period it erupted on the surface discharging huge quantities of oxygen into the atmosphere and triggering the biological Cambrian explosion of multiple life forms. The huge amounts of almost pure iron ore found sitting immediately on top of the Pre-Cambrian rock that forms Labrador's Laurentian shield is highly suggestive. There the iron ore deposit is many hundred of metres thick extending over several hundred square miles.

It has been suggested that the earth's heat is from radioactive decay, mostly from Uranium Uranium 238 decays to lead, atomic weight 207.19 producing 8 units of helium nuclei 4.0026 en route. There is little spare mass to produce heat energy. U238's very rare isotope U 235 produces one proton's worth of energy in its decay. 235 Kg of the isotope would produce 9×10^{16} Joules eventually. But its half-life is in billions of years unless accelerated by neutron bombardment. It would take more than a million tons of uranium ore to produce this 9×10^{16} Joules over a ten million-year period. It would take an earth's mass of uranium ore to generate 10^{32} Joules for the earth's heat. There simply is not enough uranium ore in the world.

Oxygen arises from the fusion of helium, normally within a helium-burning star, but also in the depths of giant superstars before they undergo Type 2 supernova detonation. The subsequent supernova would release that oxygen into space as well as generate more oxygen. Within the universe oxygen is the third most common element, after helium and hydrogen. Its relative abundance is 0.6% of that of helium and a tiny fraction of that of hydrogen, but the bulk of those two gases are locked up in stars. Oxygen is sixty times more abundant than carbon. That is if all the carbon combined with the oxygen (and go on to form frozen CO_2 comets) there would still be a gross excess of oxygen. Much would probably combine with hydrogen to form dirty snowball comets. But clearly some of the oxygen would be propelled sufficiently far from the heat of the supernova to be too cold to react with any other element.

The possibility thus exists that following the detonation of the parent supernova there was an expanding cloud of debris very rich in oxygen. The debris settled around the sun. (Whether the sun formed from a fragment of the outer shell of the supernova or was a young companion star to the giant that formed the supernova is open for debate.) When the fragments coalesced and formed into a disc around the sun, the oxygen accompanied them. As the rocky proto-planets grew in size so the oxygen was also drawn from the local area, stripping the local space of oxygen. As the fragments heated up and moulded to form planets, such as Earth, so the oxygen chemically reacted with the other elements so that the earth emerged in a ring of fragments denuded of free oxygen. The same thing occurred with the other rocky planets. Earth then underwent its evolution meanwhile oxygen from the outer reaches of that cloud that was forming the solar system was very slowly diffusing to fill the denuded space around earth. Any molecule of oxygen drifting near earth would be captured by earth's gravitational pull. Slowly, but very slowly the oxygen level would rise.

But this cannot be the whole story. The Cambrian explosion of complex life also suggests a sudden dramatic increase in environmental oxygen. The changes in the iron oxides also suggest that at sometime there was a sharp and sudden leap in oxygen concentration.

Meanwhile in the outer reaches of the solar system fragments of that original supernova were condensing. Any hydrogen combined with oxygen to form water ice asteroids or giant dirty snowballs of ice, raw carbon and frozen carbon dioxide. But the surfeit of oxygen surely means that clouds of oxygen must have formed and frozen to form oxygen snowballs. Some probably are still in the Oort clouds or Kuipers belt. Then about six hundred million years ago earth had a lucky strike.

A large oxygen asteroid hit the earth. It had a mass of about 10^{20} tons. For comparison the moon has a mass of 7×10^{19} tons. A wild or perhaps not so wild conjecture is was it responsible for smashing of a chunk ~1.5 per cent, off the earth that went on to form the moon. Or were there two massive hits by very large asteroids, one responsible for the moon and the other responsible for the planet's oxygen supply. The Cambrian explosion of life suggests that almost certainly there were two hits. Without this lucky strike life on earth was doomed to fail, as there were no renewable sources of energy to fuel the life. Once the initial dowry of hydrocarbons had been used up life would have to shut down. At the time of the asteroid impact life would only have reached the stage of forming thin sheets, perhaps two or three cells thick but anaerobic metabolism is so inefficient that the formation of larger organisms with cell differentiation is an impossibility.

All this is a somewhat convoluted way of saying that if there were intelligent life on other planets they too would have had to be in receipt of a lucky strike. But the starting conditions would have been the same, a Type 2 super nova near a young star, an initial anaerobic atmosphere, a generous endowment of hydrocarbons, a large mass of debris which including the formation, among other things, of oxygen asteroids, one of which hit the planet. The planet would have to be just the right distance from the young star and also would need to have plenty of water. The probability of all these things occurring is very small. It happened to earth or we would not be here. Given the rate at which other stars are found to have planets the odds are shortening that there may be intelligent life elsewhere in our Galaxy but the occurrence must be very rare.

The probability that all this occurred with two stars that are within fifty light years of each other is remote in the extreme. There are less than one thousand stars within fifty light years of earth. Which gets back to the probability of success of the SETI, the search for extra terrestrial intelligence, as being so remote as not to justify the use of resources.

A question of religion

A profound question is where does God fit in all this, assuming that there is a God. The immediate riposte is which or whose God. The history of religion shows that every community on earth has invoked the existence of God or Gods in some shape or form, from

Meso America, to Peloponesia, to tropical Africa, China, India, as well as the more well known locations, Egypt, Greece, Rome and Scandinavia. Characteristically all the early religions were polytheistic and had their gods fighting each other or with other gods or demigods. Their gods were not particularly interested in people, who were expected to pay tribute to them. Mankind was for their amusement. Tribute was maintained by armies of warrior priests—even up to the time of the Crusades. Anybody who disputed the authority of the priests that are the representatives of the God was to be put to death. Even now that tradition appears to exist in some parts of the globe. In the past this was much worse. The cleverest man in all Greece, Socrates, had to die for expressing a belief in monotheism. Galileo nearly died for expressing an idea that challenged the authority of the warrior priests, the Inquisition. What however is particularly noteworthy is that in all these various religions there was a profound link between religion and astronomy. Great buildings for religious purposes were built, in Peru, in Mexico (the sun and moon pyramids) in Egypt with their pyramids, and in England (Stonehenge). Whether in the mountains of Machu Pichu, the deserts of Egypt or the plain at Salisbury in Wiltshire, England, their monuments were all carefully aligned with astronomical objects.

Later emerged the concept of a single God. First propounded by the Jews, and later independently deduced by Socrates, and later still was Islam. Attention also switched from Gods who were essentially indifferent to men to the concept of a God who is interested in every individual at an individual level. It is a matter of belief, or faith by the individual—although others proselyte or even kill to ensure recruits to this or that religion, all of which seems contrary to the concept of a caring God. Hard incontrovertible objective proof does not exist. Many, on the basis of personal happenings to them during their lives, believe there is a personal God. This is particularly so when the probability of those personal happenings occurring by chance are so remote that by normal scientific protocols and standards one would regard them as not being due to chance. But there are other hints.

Possibly the most remarkable, but nowadays is so commonplace that it is taken for granted, is the emergence of compassion. This has led to the abolition of slavery, the concept of health facilities being readily available to the poorest members of society, to the granting of Aid, financial, medical, educational, to distant peoples in far away lands. Compassion for others who are not blood relations has only emerged in any quantitatively significant way in the last one thousand years. The old lazar or leper houses were established out of self-interest, to keep the infected in isolation, but out of them grew hospices and hospitals. Compassion, with one very specific but very strong exception of caring for one's kin within a group and especially caring for one's offspring, is contrary to evolution and Natural Selection since this diverts resources one might need for one's survival or further procreation. It also promotes the existence of competitors for potentially scarce resources in the future as well as competition in passing on one's genes. There is no element of enlightened self interest in saving distant people who would otherwise die, say from starvation or ill health. The growth of compassion is of religious origin, the prime example being the Good Samaritan. Its success suggests outside influence given that it conflicts directly with personal selfishness and selfish genes. It should, under the laws of Natural Selection have died out. Instead it continues to grow.

Science cannot provide proof of the existence of a God. Science is concerned with the how and not the why. Religion is concerned with the why and the purpose of life. Nature has only one single purpose, the reproduction of its genes. Everything else is subservient to that imperative, and science is the study of all factors which could influence that imperative.

Scientific logic also deduces that rare though it is, intelligent life must have emerged fairly often among the many billions of stars with their planets that exist in the many billions of galaxies that make the universe. This curiously puts a particular restraint on religion. If one believes there is a single personal God interested in everyone of us at an individual level then if, among all those billions of billions of stars and planets, there are not a number of stars with planets that have intelligent life God has been sinfully wasteful of resources. Those resources could otherwise have helped the individuals about whom, it is maintained, He cares. That by religious standards is a contradiction. That is according to religious logic there must be life in those other planets. But then a personal God would have to be interested in each individual in all those myriad of planets. Which raises a problem of communication, since many of those planets are billions of light years away, unless each planet has its own personal god in which case the concept of monotheism is destroyed. If there is a single God personally interested in each individual there must be system of transmission of complex information at velocities higher than the speed of light. And if God specifically influences certain events, even at an individual level, there must be the transmission of energy at speeds faster than light. The distances are too great for science ever to know how it is done.

There is one piece of religious writing written more than three thousand years ago which is profoundly disconcerting. It occurs in the Old Testament in the book of Job. It states, (God states) that there are rules that govern the heavens and then goes on to talk of laws of nature on earth. This was written at a time when there was no concept of science, or scientific method. Everything was considered as consequences of individual actions by God, or just happened and were not governed by specific laws of Nature. In that same text and immediately following specific references to Orion's belt, the Zodiac and the star Aldebaran, comes the suggestion of the existence of possible rules governing the stars where most stars had obvious fixed positions and distance relationships with each other. This flies in the face of contemporaneous knowledge. There was not even a concept of a heliocentric solar system. It was not until nearly two thousand years later that there gradually emerged the proof that there are laws governing nature on earth, and rules or laws governing the stars, such as gravity. That piece of religious writing is very, very puzzling, appearing as it does among people who were essentially pasturalists and for whom the information about order prevailing among the stars had little obvious relevance to their daily lives. The names given in that text are Arabic names, yet the study of astronomy in Arabian countries did not begin until well after the book of Job was written. Which raises a minor mystery in its own right.

The temptation is therefore to dismiss the existence of God as being too improbable to be true. That in turn imposes specific restraints on the Scientists. Quantum theory requires that scattered throughout the universe pairs of virtual particles pop out of space. These are virtual electrons and positrons. Rarer are the virtual protons and antiprotons. For the most part they undergo mutual annihilation releasing a photon or photons. Photons are packets of energy and this therefore conflicts with the axiom that energy cannot be created or destroyed. A particular aspect of this theory is that if the pair pop out of space very close to the event horizon of a black hole one particle could disappear into the black hole the other would remain and become a real particle. There is no way that hard concrete evidence can prove the transient existence of virtual particles. The probability of such an event is, well, too improbable to be true. Consistent logic therefore requires that if certain things are too improbable to be true they should be dismissed. That is quantum theory should be dismissed. But quantum theory is too successful in explaining many scientific phenomena to be dismissed out of hand. Logical consistency requires that if you believe in quantum theory and virtual particles, you should believe in a personal god. But when has mankind ever been consistently logical?

There is one bizarre set of facts, which defies scientific explanation. This is the extraordinary precision of the fundamental constants in nature. These are things such as the charge of the electron, the size of the Gravitational constant, the speed of light and so on. We owe our existence to these constants. This has been called the anthropomorphic or anthropic principle. Had the gravitational constant been any bigger then supernovas would not explode and the elements that make our being would not have been created. Similarly with the charge of the electron, had it been different protons would have formed into neutrons which have a short half-life before disintegrating into energy. It has been hypothesised that there could be variant universes where these constants were different. Most of these universes would be still born and so far the only successful universe among the various computer simulations is the one in which the constants are the same as those that exist in our universe. Either way the precision of the constants is not a random event. But neither was the canine tooth of the sabre-toothed tiger. The latter took billions of small evolutionary steps against a background of a particular series of environments. There is no evidence at all of any series of evolutionary steps dictated by the environment for the emergence of the various but very specific constants of nature. There was no environment at the time of the Big Bang, just amorphous empty space. This has been used by many for the proof of the existence of God. But that leads to the argument who created God. That is either the constants always existed or God always existed. There obviously will never be an objective resolution of the dispute.

Perhaps the best resolution of this schism between religion and Science is given in Philip Ball's biography of Philip Theoprastus Bombast von Hohenheim, otherwise known as Paracelsus. The book is entitled The Devil's Doctor and on p 8 he writes: "A mechanistic view of nature can be traced back to the great rationalistic scholars of the twelfth and thirteenth centuries: men like Tierry of Chartres, William of Conches and John of Salisbury, who argued that God did not run the world with His hand constantly on the levers but instead formulated rules and then let them unfold." Those rules, we now know, cover the fundamental constants such as the charge of the electron, the speed of light etc. Whilst these Christian scholars could not possibly have known of these rules the principle they enunciated when stripped of its theology is identical with the Anthropic principle.

To end this chapter on a very personal note. My God has a sense of humour, and likes to tease. Some while after the sudden death of my first wife it was suggested to me that I should marry again. I vowed no as I would never meet another woman as good as my late wife. Within forty-eight hours there was a knock on my front door and there stood a lady, a stranger, who eventually became my second wife. It took me some time, months, to realise that I had emulated Wagner's Flying Dutchman. But it did not end there. The most blatant tease occurred on the day that we got married, in late spring in the middle of England. We had agreed not to have a white wedding. On our wedding day it snowed. It was the only day for a year that it snowed. It rarely snows in the middle of England, in late spring. Why should that be on the one day we chose for our not to be white wedding it snowed? Statistically this combination of four such potentially chance events, the vow, the knock on my door, the agreement and the snow and their timing occurring to one person is remote in the extreme. It is even more remote when it is realised that that person is interested in probability theory and the quantitative consequences of cosmology. Indeed given that the odds of this combination occurring by chance are so remote it is perverse not to look for an alternative explanation. But it is a matter of belief. Or am I being an old romantic?

Chapter 10

Modernising the Big Bang

The Big Bang cosmology, critical analysis, time and temperature, the Big Bang cosmology modernised.

It is worth recapitulating and in more detail the Big Bang hypothesis and performing a detailed critical analysis of the hypothesis to show why the theory needs modernising.

The Big Bang Cosmology

The current cosmological hypothesis runs as follows:

In the beginning, about 13.7 billion years ago, there was just a singularity. It was surrounded by nothing, not even space. The singularity had a radius of one Planck length, or 4.1×10^{-35} metres. It was an extremely dense form of energy, the equivalent of almost 10^{182} degrees Kelvin, although temperature at this scale and size is meaningless. This temperature is calculated from the apparent present temperature of space as 2.7 °K in a spherical volume of radius 13.7 billion light years, to the temperature that would emerge when that same amount of energy is confined to a volume the size of the singularity.

Space and time did not exist until the singularity erupted. This eruption was the Big Bang. With the eruption came time and space. Space has been expanding ever since but time has maintained its constant rate of change. The initial expansion, up to the size of about one cubic metre, was faster than the speed of light, so enabling energy to be uniformly distributed through the enlarging singularity. This uniform distribution of energy is necessary as from this energy eventually came the galaxies that are uniformly distributed through space. This period of faster than light distribution of energy has been called inflation. Precisely when time started is questionable, but if time did not exist until after the inflation episode the description of inflation as being faster than the speed of light is also meaningless.

But thereafter space expanded at the speed of light. In doing so it cooled. It had been thought that at the end of about three minutes, actually 3 minutes and forty-six seconds, that temperature had fallen to 9×10^8 °K. However even expanding at the speed of light from an initial radius of 1 metre the resulting volume would be too small for such a relatively low temperature. Alternatively at the time of clearing (see below) the radius was 200,000 light years, the age 200,000 years and the temperature was $\sim 10^{15}$ °K.

With the energy was a force that subsequently split into four, the strong intra nuclear force, and the weak intra nuclear force, the electromagnetic force and gravity. Initially the forces were all of equal strength but over time their energy was dissipated by the volume of space until the forces became locked on or in atoms.

The expanding space was full of photons and neutrinos. Their relative density was such that any light photon would be in collision with another photon so that the universe was opaque. By the time the universe was 200,000 years old it had expanded sufficiently that the universe was now transparent. At that stage it has a radius of 200,000 light years, that is its radius was four times that of our Galaxy. The temperature had now fallen to approximately 10^{15} °K from the original 10^{182} °K. This is still too hot for atoms to exist. This stage is sometimes called the stage of clearing.

By the time the universe was almost four million years old the temperature had fallen to 45×10^6 °K. This is the highest temperature at which protons can exist. At this temperature their velocity is almost the velocity of light. Any higher temperature would have required a higher velocity but no mass can exist at such a velocity because of the time expansion due to the velocity. Meanwhile out of the energy had condensed electrons and positrons, which are simply small packets of energy. Mutual annihilation resulted releasing their energy which promptly condensed into packets reconstituting the electrons and positrons. But with each precipitation of energy into these packets there was an excess of electrons. The orgy of annihilation and re-formation continued until there were no more positrons left.

Meanwhile other packets of energy had started to precipitate as quarks of various kinds and flavours. As the temperature fell further the critical 45×10^6 °K was passed. The quarks could combine to form protons and neutrons. Antiprotons and anti neutrons also formed and there was a repeat of the orgy of cycles of mutual annihilation and reformation. Once again there was an asymmetry and so eventually only neutrons and protons existed. It was sufficiently hot for the protons to start fusing to form helium and a small quantity of other very light elements. This released some energy which raised the temperature somewhat but not to any significant extent. But with the expanding space the temperature was steadily falling and by the time it had got to below $9 \times 10^6 \,^\circ K$ it was now too cold for further fusion. By this time the universe was now 63 million years old, with a radius of 63 million light years.

The huge great clouds of hydrogen-helium mixture began to shrink under the influence of gravity, breaking up into distinct patches. As they contracted, within each cloud bits of the cloud contracted further, due to a statistical random scatter of the concentrations of the atoms. These bits were to form stars. As they shrank into themselves their centres became heated up, as their energy was concentrated into a smaller and smaller volume. The bigger the bits the quicker this occurred so that giant stars were first formed. As they heated their cores it became hot enough for hydrogen fusion and so luminous stars were formed. They were giants, each was several hundred times the mass of our sun. Naturally the centre of the protogalaxy had the greatest quantity and density of the original hydrogen-helium cloud and so these giants were concentrated at the centre of the galaxies. Both hydrogen and helium stars were formed. The helium stars proved to be unstable and exploded as type 1 supernovas forming the dust that forms a significant mass within any galaxy.

The giant superstars quickly burnt out their fuel and exploded as type 2 supernovas making even more elements that formed yet more inter stellar dust. The end result of a type 2 supernova is a black hole. With so many of these superstars near the centre of the galaxy, each exploding as giant supernovas, the remaining stars near the centre were blown outwards forming a central bulge of the galaxy. The galaxial rotation had meanwhile reduced the galaxy into a disc shape. The black holes became attracted towards each other and eventually collided to form a giant black hole at the centre of the galaxy.

But superstars also formed in other parts of the galaxy and they too went through a stage of profligate burning of their hydrogen fuel and ending as supernovas. More than half the stars in any galaxy ended up in binary systems, two stars orbiting each other. Others form triplets and even quartets. The size of the stars in any partnership had little bearing on the pairing.

One such pairing, a superstar joined with a young small star near the outer reaches of our Galaxy. The superstar eventually exploded, showering its young partner with large quantities of residue. The young star's gravitational field picked up some of this residue. The heavier elements were pulled closer to the star and settled into orbit around the star. Gravitational attraction soon saw the fragments coalescing into forming the rocky planets. Further outwards the more gaseous parts of the superstar's residue similarly coalesced into forming the giant gas planets. One intermediary planet became torn between a gas giant and the sun and when hit by a large asteroid broke into a mass of fragments. Meanwhile on the very outer fringes of the planetary system other fragments of the supernova coalesced into much smaller but still significantly sized planetissimals that remotely orbit the Sun, the Oort clouds. Thus the solar system was born.

From the distant Oort clouds the occasional asteroid would become dislodged and swoop towards the sun as a mighty comet orbiting the sun in an elongated and very narrow ellipse. Such a comet would strike any planet or moon in the way. Most of these were dirty snowballs, a complex mixture of dry ice and water ice. Other comets came from the shattered failed planet and rained in on the innermost planets. One particularly big comet hit the earth and caused it to lose about one and half percent of its mass. That mass eventually reassembled as the moon.

Within the neighbourhood of the earth all the oxygen formed by the supernova had combined chemically with other elements so that the proto-earth was denuded of free oxygen. It was however well endowed with a large quantity of simple hydrocarbons. Some of these reacted with each other and formed hydrophobic collections of progressively more and more complicated chemical compounds and so life was eventually started.

The primitive life forms existed as essentially mats of single cell thickness and existed anaerobically on the hydrocarbon dowry. Eventually photosynthesis emerged and this improved the ergonomic balance. The lifeforms could now store nutrients and metabolise them at their leisure.

There is a residue of the Big Bang and this is in the microwave radiation that surrounds everything. This is the background microwave radiation, sometimes abbreviated as BMR. This has been attributed to the after glow of that Big Bang and because space has a limited but large volume the temperature has fallen to $2.7°K$ The space also shows hot spots where the temperature is slightly above the BMR. These hot spots are attributed to patches of the energy of that initial fireball that have not come into equilibrium with the rest of space. The otherwise uniform low temperature, as shown by the longer wavelength of the microwave radiation, is due to wave stretching by the expansion of space. The red shift of distant receding galaxies is also due to the stretching effect of the expansion of space.

As to the future the current cosmological thoughts vary. Chief among the uncertainties is the natural geometry of the universe. If it is flat, that is it follows Euclidean geometry completely, then the universe will continue to expand for ever. Such an expansion is antithetical to the concept of curved space.

If on the other hand the geometry is not completely flat then gravity will reverse itself and the whole universe will collapse on itself, forming the Big Crunch or heat death. The evidence available points strongly towards a flat universe but the proofs have as yet not reached 100% certainty.

With unlimited expansion the galaxies will become further and further separated until they disappear out of sight. The driving force for the expansion has been called dark energy. As space expands it drags the galaxies with it. Space therefore has a physical attachment or adhesion to the galaxies which can be over ridden, if the gravitational pull from a neighbouring galaxy is strong enough, so that galaxies collide.

Not all the energy that was present before the appearance of quarks ended up as mass that would form the stars. More than seven times that which formed luminous stars ended up as a mysterious form of matter called dark matter. This dark matter is, as its name suggests, non-luminous. But it effects can be seen by the anomalous velocities of the outer stars in the spiral galaxies. They are moving too fast unless the gravitational pull from the centre of the galaxy has been augmented by something that emits gravitational energy. It was this that has led to the concept of dark matter.

Eventually all the stars in all the galaxies will have used up all their nuclear fuel and all that will be left is very slowly evaporating black holes and the clinker of dead galaxies. This will take many thousand times the present age of the universe.

Well before this our sun would have used up most of its hydrogen fuel and would begin to contract down to a white dwarf. It would blow off its outer shell of unfused hydrogen forming a red giant that would engulf the earth, heating it to a surface temperature of several thousand degrees. Incinerating it. This will not occur for several million, if not a billion, years.

Critical analysis

Although the Big bang hypothesis is broadly in agreement with most physical science yet there are a number of serious problems, which undermine the whole hypothesis.

(i) First and foremost is that the expanding space concept assumes that space is a physical substance rather being simply an absence of anything. Not only that by expanding physically it breaches the constancy of the velocity of light. The latter is a fundamental constant that is incorporated into many other fundamental values, such as Planck constant of length, the Planck constant of time, and indeed the whole of relativity theory. There is no proof that space has expanded. However general relativity theory and its conclusion about a space time continuum requires that if one expands the other must as well. But in this Big Bang hypothesis time has not expanded. It has increased in quantity but the second has the same period as it had just after the inflationary period.

(ii) If space is limited then what happens to the photons emitted by galaxies that are close to the boundary? Are the reflected and if so what reflects them? If they are absorbed this implies that the space can absorb light. It also implies that the axiom that energy cannot be created or destroyed is also wrong.

(iii) A finite volume of space contradicts general relativity theory of a curved space-time.

(iv) The cosmology assumes that protons have an almost infinite life. Modern physicists have theorised that the life of the proton is about thirty or so billion years. This eliminates both the Big Crunch hypothesis and also the infinite continued expansion of space.

(iv) As it stands the Big Bang theory neglects the acceleration factor in the expansion of space. The force applied is an accelerating force, and so at some point the mass that makes the galaxies must reach the speed of light at which they would cease to exist as mass.

(v) It does not explain how the expansion force is controlled, or its mechanism of release, or where the energy required sustaining and expressing that force is stored. This energy is often called dark energy.

(vi) Dark matter is presumed to exert a gravitational force effect. It should therefore attract to itself more dark matter and so go on to develop into gigantic superstars of dark matter within a galaxy. Such conglomerations of such intense gravitational fields should be manifest in serious disturbances of the orbits around the central point of the galaxy. They would act as million or even billion solar mass gravitational objects A multitude of such objects would disrupt the entire geometry of the galaxy and this would be evident. To postulate that the dark matter can exert a gravitational force but that the force does not affect other bits of dark matter is being very selective and reminiscent of the epicycles postulated by past astronomers. Such conglomerations of matter should heat up due to gravitational heating, emit microwaves and explode as supernovas. This is not seen.

(vii) If there is an external force pushing the galaxies at ever-increasing velocities then the structure of the galaxies is not sustainable. Galaxies consist of a mixture of the most dense, that is black holes and neutron stars to the least dense, inter stellar gas and dust. Any external pushing force would have a stronger effect on the least massive of the objects within a galaxy, such as the dust. The reflected dusty glare seen around the centre of reasonably close galaxies suggest that the galaxial structure be not being disturbed by the expansionist drive.

(viii) It does not explain what is glowing in the hot spots. They cannot be pure energy since energy travels at the speed of light and would have dissipated long ago.

An entirely different set of problems relate to time

(ix) The Sloan galaxies are too far away. There has not been enough time for them to get that far. Postulating that space expanded in between and has carried the galaxies that distance, or that space is expanding whilst the light is in transit requires entirely new laws of physics

(x) The attribution of the red shift to the expansion of space and so stretching the waves does not accord with astronomical observation. In particular the observations that the outer stars of a large galaxy are travelling faster than predicted. If the red shift is due to the expansion of space between the galaxy and Earth all the stars in that galaxy should have the same red shift. The facts point otherwise. Indeed the difference in red shift between one side of the galaxy and the other has been successfully used to quantity the value of Hubble's constant. Such a mechanism proscribes the expansion of space hypothesis and its stretching of the microwaves.

(xi) It does not explain how the globular clusters are apparently older than the age of our Galaxy.

(xii) Perhaps the biggest difficulty of all is the lack of time. A number of the Sloan galaxies have been identified by the explosions of supernovas, including type 1 supernovas within any particular galaxy. Those galaxies are over ten billion light years away. Supernovas reflect the death of a star. Even the biggest stars have lives that exceed a billion years, whilst type 1 supernovas are the end result of a star that does not exceed eight solar masses with a corresponding life of at least five billion years. There simply has not been enough time for the giant gas clouds of the protogalaxy to condense to form small clouds. The clouds had radii that are at least fifty thousand light years in length, and contained enough material to condense to the density required to form stars and initiate internal hydrogen fusion. There may be a caveat to this query in that the type 1 supernovas could have come from pure helium stars that formed out of the primeval hydrogen helium mixture. But this would mean all of them were primary helium stars. But no such excuse exists for the type 2 supernovas. One contribution to the lack of time concept is that it was too hot for protons to form until the universe had expanded so that its radius was just over three and half million light years. This would have delayed the onset on condensation of the giant primordial gas clouds.

(xiii) Perhaps the greatest criticism of all is that the Big Bang theory ignores the effect of gravity. With so much mass around, the original singularity or its immediate development should have resembled an enormous black hole. The gravitational effects of such a hole would have prevented anything inside it from exploding. The alternative is that gravity did not develop until mass formed which in turn was some time after the Big Bang. If that is the case gravity is not a fundamental force of nature. G is not a fundamental constant. But G is incorporated into a number of fundamental values, such as the units of Planck length, time etc.

Is Time temperature dependent?

When the expansion constant H_T is unconstrained by intense gravitational forces, the resultant changes in radius will indeed reflect the expansion of time. There is a suggestion that time itself is temperature dependent. Immediately in the vacuum over a Bose-Einstein condensate the velocity of light is slowed to a few metres per second. Since the velocity of light in a vacuum is constant in all time frames the observed slow velocity of light means that time itself has slowed. The temperature of the condensate is a small fraction above absolute zero temperature. A similar situation occurs in black holes. There the temperature is a whisker above absolute zero and the pace of time almost infinitely slowed. Conversely when the temperature of the universe was very hot, at or above the temperature that energy could condense into nucleons, the pace of time was extremely fast. That is it was more than a million times faster that present earth time. The inference is that at these extreme temperatures time speeds up. Then as expansion takes place so cooling occurs, and this slows time. The expansion causes both the radius to increase and the temperature to fall. The inference continues as every time the temperature fell by a factor of eight so the period of the second doubled. This is because doubling the radius increases the volume by a factor of eight and temperature reflects the mean density of heat photons within any volume. It is a coherent and intriguing explanation and could be the background that doubling the radius of the universe causes time to slow by a factor of two.

The other aspect is that it has been shown that time is inversely related to mass. But mass is energy. Energy reflects temperature. The inference is that when the energy is very low, the temperature is very low. The inverse of this is time which is very large. Conversely when the energy or temperature is very high the time is very small or very short. But considerably more research is needed to confirm any mathematical relationship between temperature and the velocity of light or time. Nevertheless it is an intriguing possibility.

The Expansion of time and modernising the Big Bang Cosmology

Many of the difficulties described above disappear if one takes into account that time is expanding and has been expanding since the fireball had expanded .to be bigger than its event boundary. Every time the radius doubles the pace of time halves or the period of the second expands to twice its previous size. It is important to stress that this is not a causal relationship as far, as can be determined, although the temperature suggestion above suggests that there might be a link.

The subsequent figures are based on the assumption that the visible universe has a radius of 13.7 billion light years. This assumes the NASA figure for the age of the universe. Although all the discussion has talked about the visible universe the furthest object seen has been a quasar that has been positioned on Gott and Juric's map, as being at a distance of 10^{193} earth radii. This corresponds to a distance of 13.44 billion light years. Clearly, with improved telescopic technology this is unlikely to be the limit. For convenience therefore it was therefore decided to retain the 13.7 billion figures for all calculations such as the doubling number and so the age of the universe in cosmological time. One other assumption is that there is a background microwave temperature that permeates all space including extra galactic space but it may not be the 2.7 °K.

The modernised Big bang cosmology then is as follows:

There was originally a ball of energy sitting deep in a vast sea of space. How big it was is impossible to say. In that fireball was all the energy required to make all the mass of the universe. The mass of the visible universe is currently 3.33 x 10^{53} kg. (See technical note 1) The corresponding energy therefore is 3 x 10^{72} Joules. But possibly, due to the effect of the speed of time on mass, it was almost two hundred million times more than that, that is 6 x 10^{80} Joules.

The temperature was extremely high, although again it is impossible to deduce how high. Within the fireball were the five, possibly six, fundamental forces of nature, namely the forces that were destined to become the strong and weak intra nuclear forces, the electromagnetic force, gravity, and the expansion force. A sixth force has also been propounded by the atomic physicist, Burkhard Heim, but its role is unclear. All these primary forces were of equal strength. The forces were carried by their appropriate particles, photons, gluons, accelerons, and gravitons. But such was the effect of gravity that the force particles, other than the gravitons, could not escape. But just as gravitons can escape from a black hole in the centre of a galaxy, so the gravitons could radiate out into the deep space. The other energy particles were trapped within the Schwartzchild radius or event horizon The event horizon is where the acceleration due to gravity equals c the velocity of light. The question therefore arises what would be the Schwartzchild radius if the entire mass of the universe formed a black hole. Within a black hole time is at a virtual standstill. An important consequence is that time could only start when the radius of the universe was bigger than its Schwartzchild radius. This was, at a minimum, approximately 29 light years from the centre of the fireball (but possibly the radius was 22,000 light years, see technical note 2). For comparison our Galaxy has a radius of 50,000 light years approximately

Within the fireball, because of the intensity of the gravitational force, time did not exist. Time could only start when the other energy particles could escape from the fireball. The fireball, although massive, had a limited size. This was determined by gravity. But the gravitons were radiating out into the outer space. Eventually the loss of gravitons crated an imbalance so that the fireball began to swell as a result from pressure of the accelerons. There then came a point where the radius exceeded the radius of the event horizon. Some of the energy particles were then able to escape from within the fireball.

But immediately around the fireball, gravity was so strong that the energy particles were refracted to form a tight orbit around the fireball, that is a gravitational lens effect. (Another interpretation is that space, or the space time continuum, was distorted so much by gravity that photons and other energy particles formed a circular orbit around the fireball. This interpretation implies that space has substance which may not be true). The energy particles were initially confined into a narrow blanket around the fireball. Their density was so great that they could and did combine to form bigger particles.

First the photons combined to form electrons and positrons. Gravity confined them into the narrow blanket and they readily collided releasing their photons, which promptly combined again. This orgy of combining and destructive collisions continued but there was an imbalance. More electrons were produced than positrons. The end result was the elimination of all the positrons. Meanwhile the accelerons was combining with the gluons and the gravitons to form a whole family of sub atomic particles, which combined to form quarks. They in turn, combined to form protons. Some of the protons combined with electrons to produce neutrons. It would seem that just as electrons and positrons evolved so also evolved the protons and antiprotons. They too went into mutual annihilation and reforming and again the same imbalance prevailed, until there were no more antiprotons.

Free neutrons have a very short life that is measured in minutes. They have to combine in some way with protons; otherwise they break up into their constituent energy particles. Some 30% of the initial population of protons (hydrogen nuclei) combined with electrons to form neutrons. Another 30% of the initial population of protons combined with the neutrons to form helium. The end result was the 27% helium concentration that is a feature of interstellar gas. Only approximately 40% of the original stock of protons remained and it was this residue which was form the 15+ billions of galaxies

Meanwhile the fireball's event horizon was still contracting deeper and deeper into the fireball. More and more energy particles emerged and went through their respective orgies of combining and mutual annihilation. The initial emerging protons and helium nuclei were massive, more than a quarter of a million times their present mass. This was because the pace of time was so high.

As the event horizon retreated so there were an increase in the distance between the surface of the event horizon and the

243

original protons and helium nuclei in the outer parts of the blanket around the fireball. This weakened the gravitational hold both on time and on these masses. Time meanwhile had started and the pace of time quickly rose to become 2^{19} faster than the present earth time. The second was correspondingly relatively brief. But it was as cosmological time. As the distance gap between the event horizon and the outermost protons increased further, that is the radius increased, so time began to slow. This caused these outer protons to shed some of their excess mass as energy. The energy released was in the form of both gravity and acceleration. That is the atomic nuclei were able to develop a significant gravitational force of their own. This interacted with the gravity from the fireball and caused the particles to stay fairly close to the fireball. That is the expansion of the mass component of the universe was very significantly slowed. But the photons that had not been caught up in the nucleo-synthesis could escape and radiate out into space. That is the radius of the enlarging heat and light carrying universe, as defined by the photons it was carrying, was able to expand at the speed of light. With each doubling of this radius so time slowed by a factor of two. Equally each doubling of the radius caused the volume to increase eight-fold and the resulting expansion coupled with the escape of some of the heat photons caused the temperature in the outer parts of the blanket around that fireball to drop to below that required for hydrogen fusion. This was a lucky escape. There was enough hydrogen left to form galaxies that could last for billions of years. The helium stars, and they must have formed, were too unstable to last for so long. They would have exploded as supernovas.

Deep within the fireball's surrounding blanket this history was being repeated again and again until eventually there was no fireball left. Now the remainder of the protons and helium nuclei could do their thing. Join up as clouds, and condense to form stars. These later nuclei were less massive than the very first born because time was slowing significantly and because of the inverse relationship between time and mass.

Meanwhile the massive hydrogen and helium nuclei were picking up electrons and gravitationally collecting into dense clouds. By the end of the first doubling period they would have shed half their mass as energy. This included gravitational energy, which facilitated their condensation into dense clouds, as well as acceleration energy. But the gravitational pull of the original fireball would still be sufficiently strong to prevent the clouds from escaping. The acceleration energy would simply accelerate the velocity of the clouds in their circular orbits around the fireball. Gradually however the orbits around the fireball increased in radius Meanwhile the density of energy particles in this surrounding blanket was rising as more and more particles escaped from the fireball's event horizon, and formed more and more nucleons and electrons. Gravitons were being discharged into far space. Slowly, but how slowly is impossible to say, the blanket around the fireball was so thick that the fireball's very slowly diminishing gravitational energy could no longer hold them close and until they could escape into distant space. The effect was similar to that of a Catherine wheel firework. This was a slow process as layer after layer of the blanket escaped into space. The net result was that space was uniformly seeded with galaxies with the most distant being the oldest.

It is impossible to say how long all this process took, as there are two complicating factors. These are the acceleration constant and the half-life of a proton

The acceleration constant of 4.86 x 10^{-10} metres/sec/sec of cosmological time means that the proto-galaxies would have reached the speed of light in 20 billion years or two doubling units (in earth time units this would be just over ten billion years). However relativistic effects come into play. As the galaxies approach the speed of light their time slows. At a velocity of 0.9c their time dilatation factor would be a little over two. At 0.95c the time is dilated more than three fold. At 0.99c time is seven times as long. That is relative to earth time the galaxies are slowing up considerably. But to the galaxies they are still accelerating at 4.86 x 10^{-10} metres per second per second, but the second is now prolonged due to relativistic time dilatation. Their velocities, as far as the galaxies are concerned, are still approaching c and more significantly their red shifts would reflect this. Because of the relativistic time dilatation their protons would not have reached their expiry dates. If this occurred whilst the nucleons were in a circular orbit around the fireball the nucleons would have lost all their mass as energy. There then would have been a repetition of the scenario of when protons and antiprotons formed and mutually annihilated each other releasing energy only for that energy to reform as nucleons. But the nucleons would have had a slightly smaller mass, as time would have slowed somewhat. The difference in mass would be due to the energy spent in accelerating the previous generation of nucleons. If the clouds of nucleons were entrapped in the fireball's gravitational field for several doubling periods then the earliest ones would also have disintegrated as energy.

As will be seen later (See technical note 2) the cosmological age of the universe is ~19 doubling periods, that is well in excess of the life of a proton. (approximately four doubling periods, see later). If some of the earlier nucleon clouds did escape from the fireball's gravitational clasp then they would have formed galaxies which are now in extended decline as the crawl towards the edge of the universe, travelling very fast in their own extremely slowed time. Extreme far space could have a host of such galaxies but as they are so dim and gravitationally emaciated there is no way of knowing of their existence.

One noticeable thing is that the light from these very old and distant galaxies, the rate of emission of photons from the stars within the galaxies, would be dimmed in proportion to the extent of their time dilatation. This dimming is in addition to the inverse square law dimming due to their great distances.

Once the great clouds of hydrogen and helium mixture got away from the fireball their evolution into spiral galaxies etc. is as described earlier in this chapter in the Big Bang theory.

The significant thing in this analysis is that the Big Bang was an extremely slow process and not something that occurred as a sudden instant. Although time initially was very fast relative to earth time it was not that fast to justify describing it as a massive explosion. This analysis though does provide an explanation for the apparent distance paradox that the Sloan galaxies are too far away to have reached their positions and shed their light back to us within thirteen billion years.

It also accounts for the globular clusters being apparently older than the Galaxy. If they were two doubling periods old then in cosmological time they would be twenty billion years old. But in earth time extrapolated backwards this mounts to a little over ten billion years. Their physics and chemical processes would be following cosmological time.

There is one other odd coincidence this is the life of a proton. A proton is 1.836×10^3 more massive than an electron. Because of the expansion of time currently protons are losing mass at the rate of 4.33×10^{-18} of the mass of the proton per second. If this rate of loss continued unchanged it becomes a simple matter of arithmetic to calculate when the proton's mass would equal that of an electron. This figure is 39.3 billion years, or four doubling periods. When this stage is reached the proton will have been reduced to the mass of a positron and mutual annihilation with electrons would or could ensue. There is a remarkable coincidence in this figure. Atomic scientists using entirely different methods have calculated that the life of the proton is 25-40 billion years. It is thus possible that their techniques of determining this figure inadvertently imply a gradual reduction in the mass of the proton, or else they have observed or deduced that over time that the mass of a proton does lessen. Could it be that they have stumbled on the same process without recognising its cause? It is an intriguing question or else a very remarkable coincidence.

But long before that the strong force and the weak intra nuclear force will have lost lots of their energy. Nuclei containing many nucleons will therefore start to fall apart. This process is not dissimilar to that which very massive atomic nucleus face and which results in radioactive decay. It is emphasised that these two may not be exactly the same not least because of the variations in half-life between very large atoms of reasonably similar atomic weight and atomic number.

The future

Galaxies are accelerating and will eventually reach the speed of light. Meanwhile the light carrying visible universe is also expanding at the velocity of light. Galaxies will never be able to reach the edge of the universe simply because of that initial delay is getting away from the fireball together with the delay inherent in building up speed to reach the approaching near light speed their internal time slows reducing their acceleration. But at this high speed they lose mass. At 0.99c they have lost 85% of their mass. The time zone of atrophying galaxies, as they are now seen. Clearly they had their own history before then.

A different problem arises from the acceleration of galaxies as they approach the limit of the visible universe. Because of the speed that they eventually reach they will experience relativity induced time slowing. The subsequent development depends upon one of two perspectives, each with its own time frame.

The first is from the perspective of an observer on the fast moving galaxy. Since acceleration (from the accelerons) is constant relative to the time frame of the galaxy in question, the observer will still feel the same acceleration. If the galaxy is approaching light speed, the relativity induced time slowing will not be apparent to the observer. But to the external universe the galaxy will be slowing and effectively the acceleration energy will be very feeble as its power is spread over an ever-expanding time. There will also be relativity-induced expansion or slowing of time that will increase the gravitational energy output of all stars including neutron stars.

This would cause them to behave like black holes. There will be an increase in the rate of quasar formation. The increased gravity will cause individual stars to contract more tightly against their radiation pressure.

From the perspective of a second distant observer there will be a number of differences. Light, whether from a supernova or even simple starlight will be dimmer as the number of photons emitted will be constant relative to the slowed time at the source of the light. Brightness depends upon the number of photons per second sensed by the distant observer's light detector. Interestingly the red shift will still indicate that the galaxy is travelling at close to light speed, as the imprinting of the changes in the spectrum (i.e. the red shift) is dependant upon the velocity as experienced by the light emitter in its time frame.) Distance as gauged by the inverse square law using the light from a supernova will have to be corrected because of this time induced dimming.

One very interesting question is how far from the fast moving stars does the zone of time slowing extend? Clearly it must extend some distance. Thus within a fast moving rocket, moving close to light speed, light travels more slowly as would be judged by a distant observer. That light is travelling in space enclosed by the rocket. If the rocket had an open frame that light would still be travelling slowly. That light may not actually be touching any part of the moving rocket. That is there must be a zone of time slowing around a fast moving object, just as there is in the vacuum above a Bose Einstein condensate. Applying this to a whole galaxy moving at near light speed raises the possibility that the space within the galaxy around a fast moving star and perhaps a little beyond, is a zone of slowed time. But a beam

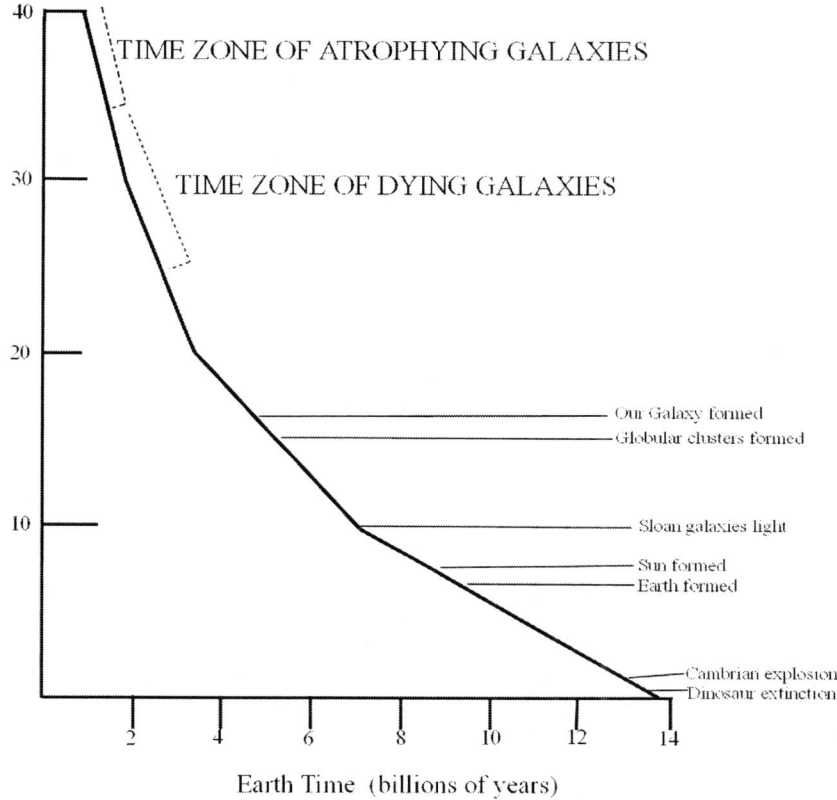

Figure 10.1 Earth time –v- cosmological time. Cosmological events adhere to cosmological time. Thus the light from the Sloan galaxies that we now receive was emitted some ten billion years ago in cosmological time. Similarly the globular clusters formed about 14 billion years ago, but this was also in cosmological time.

of light is refracted when entering a zone of slowed time, as evidenced by a beam of light from a distant star grazing the surface of the sun, where gravity has induced some time slowing. This refraction effect is discussed more fully in Chapter 6. Could this time slowing be the source of the refraction attributed to gravitational lens. The argument is that light from a very far distant galaxy is passing around or through a galaxy that is near the end of its life. That galaxy is travelling at speeds close to light and so experiencing intense dimming of its own light and creating a zone of time slowing. As the light from the very distant galaxy passes through, or close by, it is refracted creating the lens effect. It could account for the relative rarity of so called gravitational lens and why they are so distant. It makes an interesting speculation that a gravitational lens is really a time lens.

The end result is that all galaxies will be very distant but will be crawling along ever outwards taking longer and longer to make significant progress, and that will be unending. The universe will go on for ever. Long before that our sun will have swollen into a red giant and engulfed the earth incinerating everything.

Or would it? With the acceleration induced time slowing mass will have been shed. This will be mainly from the nuclei of the individual atoms. They could be reduced to the size of a positron. Their circulating electrons could then indulge in mutual annihilation. Protons do have a shelf life and when that is reached there must be some mechanism of coping with the charge each proton carries.

Technical note 1.

The mass of the universe: The mass of the Universe has been put at 10^{80} nucleons. One gram molecular weight of hydrogen has a mass of just over two grams. Avigadro's hypothesis states that one gram molecular weight of a gas has 6×10^{23} nucleons. Molecular hydrogen consists essentially of two protons. It follows that 1 gram of proton consists of 3×10^{23} nucleons. If there are 10^{80} nucleons in the universe then the current mass of the universe is 3.33×10^{53} kilograms.

Technical note 2.

The event horizon: The radius defining the event horizon of a black hole can be obtained from

(Equation 10,1) $\qquad$ Radius $= (GM/c)^{0.5}$

where G is the gravitational constant, M is the mass in kg, c is the velocity of light in metres per second, then the radius is in metres.

This is a simple variation of the equation that defines the acceleration due to gravity between two masses of which one mass is 1 kg. The substitution of c for g is because once the acceleration reaches c then can be no more acceleration. It is apparent that the dimensional units change from metres per second per second to merely metres per second. That is one of the time elements has disappeared. This equates with relativity's hypothesis that the time disappears when travelling at the speed of light, and consequentially time does not exist in a black hole.

Applying this to the current mass of 3.33×10^{53} kg gives an event horizon of ~30 light years. The current mass is derived from the postulate that there are 10^{80} nucleons in the universe. There was nothing outside this event horizon radius, just empty space. But time was much faster then and so mass must have been much bigger. That is the ~30 light years is an under estimate The original radius has now become 13.7 billion light years (the radius of the observable universe). It can be shown that if the current mass was increased by a factor of 2^{19} then the original mass was 1.74x 10^{59} kg and the event horizon had a radius of ~21,000 light years. Such an event horizon if doubled nineteen times would result in a radius of ~13.7 x 10^9 light years. The 2^{19} number is the only number that does not provide a discrepancy between the value of the present mass and the radius of that event horizon, and the current radius.

That is the original radius doubled itself nineteen times to become the present radius. Each doubling of the radius doubles the length of time and so halves the mass. The original mass was therefore 2^{19} times greater than the present mass. That original mass defines the original radius, which was the event horizon. As has already been shown each doubling unit corresponds to ~9.75 billion years of cosmological time. That is the age of the universe is 19 x 9.75 or ~185 billions years in cosmological time. And the Big Bang had to have a radius of almost 21,000 light years before anything could escape its gravity. One consequence is that our universe has gravitons extending beyond 185 billion light years and it may have some heat and light photons going as far as that distance.

Chapter 11

Epilogue

It is perhaps worthwhile to review what these studies have shown and consider the requirements of alternative proposals. First there was an attempt to understand the physics behind the Hubble constant. A geometrical analysis quickly showed that the constant could not be constant over of the whole radius of the visible universe. It is one of life's ironies that Einstein towards the end of his life was convinced that a geometrical approach would be the key to understanding the expansion of the universe. His hunch was right but he was unable to prove it. This was because he did not have the data from the far distant supernovas. It was the more recent availability of this data, which provided the necessary proof. The overall match of this data, the Hubble value at different very far distances, to the theoretical prediction was the satisfying proof.

But the Hubble constant did not make sense in terms of conventional physics. It resulted in a dimensionless number divided by time. Substituting the denominator distance in the Hubble equation with time for light to travel that distance showed that what was being described was an acceleration force. The value of this acceleration matched that found for the acceleration which had been attributed to dark matter. It also pointed to the expansion of the universe, the aberrant acceleration of stars on the edge of galaxies and the inexplicable acceleration of the Pioneer probes. The unexpected was that three apparently very different problems had the same solution. What had been found appeared to be a new fundamental constant, in the same category as the other four fundamental constants of nature, the strong and weak intra nuclear forces, the electromagnetic force and gravity. And like gravity this newly identified force was proportionate to the mass being accelerated, be it a galaxy, a star or an interplanetary probe. Like gravity it arose from the mass being accelerated, but it is an expansionist force whereas gravity is a contractual force. The existence of this force had been predicted on theoretical grounds. The identification of this force did away for the need to postulate some exotic form of dark matter.

But if it was a fundamental force it had to be constant throughout the universe. The equation describing the force showed that it was time dependent. The major step was identifying that the force only became consistent when time was expanding. Once this was put in place everything became coherent in physical terms.

But the effect of a slowing pace of time or time expansion had significant repercussions. The pendulum equation showed that expanding time reduced mass. This was confirmed in the quantum equation of the energy content of a quantum relative to its frequency It was further confirmed from the Special relativity theory..

The interesting speculation then emerges that since reducing the energy content per unit volume reduces the temperature the converse could apply. Reducing the temperature reduces the energy content and so slows time. But the reduction of the universe's energy content per unit volume is a consequence of the expansion of the universe and that is controlled by the cube of c the velocity of light. It is the limitation of c which determines the orderly slowing of time throughout the universe.

Shedding mass liberates energy. Suddenly here was an explanation of the source of the enormous energy required by gravity to do such things as controlling the orbits of planets and even inducing galaxies to collide. It also accounted for the source of the much greater energy requirement to expand the visible universe. It got rid of the need for any so called dark energy arising out of space.

But the repercussions did not end there. The expansion of time that is the result of very high velocities resulted in the conversion of mass to energy in the sun and all other hot stars. It led to a highly efficient but very simple type of thermostatic control of the temperature of these stars and so explained the remarkable thermal stability of the sun. This in turn allowed life on earth to evolve. We owe our very existence to this effect.

The expansion of time also resolved another mysteries, such as the age of the globular clusters in our own galaxy. Another mystery solved was how the Sloan galaxies could be so far away. The universe operates on a different system of time. Time has been slowing or expanding exponentially rather than progressing at an unchanging constant rate. The present rate of change is infinitesimally small but was not so at the time of the origin of the universe.

This in turn led to a re-evaluation of the Big Bang theory. Clearly the universe using its own system of time was much older than predicted by the Big Bang. That theory also took no account of the effects of gravity. But gravity as a fundamental constant has existed since time began. But gravity slows time. That is effectively time did not, could not start until whatever was the original fireball had shed enough gravitational energy for some of that other energy in the fireball to escape and eventually condense to form mass and so galaxies.

The sheer length of time and the fact that the edge of what was to become the visible universe was moving at the speed of light or close to it, meant that the background microwave radiation energy could not be a relic of any so called Big Bang. There is no reflective surface to cause the radiation to travel backwards towards its source. The expansion of time and its effect on mass suggested that this background microwave radiation could well arise from loss of mass from the dust that is close to our solar system. That dust in turn arose from the supernova that spawned our planets. The hot spots could simply be patches of more dense dust. Such an explanation could account for the very large angular size of some of the hot spots.

Another effect that emerged was the interaction of slowing time and gravity and its effect on passing photons. The resulting refraction accounted for the apparent curvature of a beam of light passing the sun and seen during a solar eclipse. There was no need to postulate that the presence of a large mass induces a curvature of space, or a time-space continuum, whatever that may be. That is the mathematics of general relativity theory that predicted that light should follow a curved path when close to a massive body could have another simpler explanation. Occam's razor implied that the simpler explanation was to be preferred.

This refraction effect could only be accounted for if energy, when in waveform, has volume. This is something the refraction gratings have shown for a long time. This volume effect means that postulating that the universe arose from a singularity of extremely high density could not be true. The singularity is too small. We cannot know the original size of the nascent universe, nor how long it existed, before it reached a size that energy was able to escape gravity's maw.

The two key discoveries of this work are that there is a fifth fundamental force of nature, which is responsible for the expansion of the universe, and that time is expanding or slowing. It was this time slowing which was responsible for the emergence of the universe. It is this same time slowing which will lead to the extinction of the universe. The extinction will take many billions of years to come into full effect. But as galaxies race away, losing mass to provide energy for their velocity and gravity they evaporate away. Long before then an expanding sun will have costumed then the earth as it inflates to form a red giant. Its fate is to become a helium star and eventually a neutron star. This neutron star will become a pulsar for other beings in other stars in our galaxy to ponder over and wonder if life ever existed on the outer planets still circulating that pulsar.

Finally was the resolution of the two greatest problems of modern science. Why the Pioneer space probes were accelerating. and is there such a thing as dark matter. It was shown that the acceleration force neatly accounted for the data involved in both problems.

What is remarkable about this whole slowing time hypothesis is how every part so neatly dovetails into the next part. There seem to be no ugly gaps in inexplicability. Shibboleths such as epicycles, celestial spheres, dark matter, dark energy are shed revealing an edifice that can be admired for its simplicity and beauty.

It is often said that truth is beauty. Beauty is in the eye of the beholder. To Aristotle and Ptolemy their constructions of the universe with its epicycles were beautiful relative to the available knowledge. But it had to go, to be replaced by the theory of great celestial spheres. This was beautiful in the eyes of the poet Dante. But as knowledge increased his near contemporaries Copernicus and Galileo created their construction of a heliocentric universe that was clearly beautiful to them. The heavenly spheres had to go. As knowledge further increased the concept of a heliocentric universe had to go. The universe was bigger than first thought. The Big Bang theory emerged and it too was and is satisfyingly beautiful in the eyes of its originators. But as more knowledge is gained so it too has to go. Now what is presented is the time slowing hypothesis. Such is the nature of scientific evolution that as more knowledge is acquired so it too, almost inevitably, will be replaced. The replacement too will be beautiful in the eyes of its originator. Clearly truth has many layers.

Not all the questions have been solved. We do not know exactly what energy is, nor time, although we can make good use of both. All we know is that energy has volume and since time is inversely related to mass which also has volume, and mass is energy in a concentrated form, time is inversely related to energy.

It is also inevitable too that there will be the deny-ers, as there were in Galileo's time but at least your author's life is not at risk. It is incumbent on the gainsayers to show by science and not by rhetoric where any errors lie. They will also have to provide some explanation for the various paradoxes, the globular clusters paradox, the distance paradox, as well as accounting for why so called dark matter does not disturb the geometry of the spiral galaxies even though the postulate is that this matter is said to account for up to nine times the mass of a galaxy. The fuel source of both gravity and the energy required for the expansion of the visible universe will have to be explained as well as the factors behind the control of the release of that energy. Probably the most difficult is to explain where the energy comes from that lifts the seas to create the tidal changes of sea level and how it gets there. Other explanations are needed for the precise long term stability of the surface temperature of the sun despite it fusing more than a million tons of hydrogen per second. And the same applies for all other stars. But if this study provokes further study and research the author will feel he has achieved his aim, to make people think

Perhaps the most challenging outcome could be determining whether the proposal of increasing the velocity of hydrogen atoms to near speeds required for fusion produces a surfeit of useful energy. Then we would be imitating the sun.

Another realisation has been that Relativity, or rather Special Relativity is simply a comparison of events set or occurring in different time frames. Nothing else. Postulating about observers in different locations and then referring to "relative to the observer" merely serves to confuse the issue. A high speed rocket manned by its own computer will still experience time slowing within that rocket but there will not be any observers.

Newton is said to have stated that he stood on the shoulders of giants. It has been the author's privilege to peep over the shoulders of those giants. Contemplating the development of the universe and how everything dovetails together so efficiently like a magnificent well-crafted piece of sculptured engineering, what he saw was beauty. Eureka!

Appendix

Some Useful numbers

Constants:

c. The Speed of light 2.9979×10^8 m/s/s rounded to 3×10^8 m/s/s

G. The gravitational constant 6.672×10^{-11} Nm^2kg^{-2}

H_o. The Hubble distance constant ~50 km/sec/M.parsec

H_T. The Hubble time or acceleration constant (as defined in this book)

$$4.66 \times 10^{-10} \text{ m/sec/sec}$$

NA. Avogadro's constant or number, the number of particles in a gram molecular weight 6.022×10^{23}

Masses and diameters

	Mass (kg)	Diameters (km)
Moon	7.350×10^{22}	3476
Earth	5.976×10^{24}	12756
Sun	1.989×10^{30}	1392000 (4.6 l.secs)
Galaxy	2×10^{39}	100,000 light years
Visible universe	$\sim 10^{54}$	$>2 \times 10^{10}$ l.years

Time

Age of the universe (Earth Time)	~13.7 billion years
Age of universe cosmological time	~185 billion years
Doubling time cosmological time	~9.75 billion years

Velocity

Sound in air   331 +0.6t m/sec where t is degrees Celsius

Sound in hydrogen   Approximately 4m/sec/°K

Primary equations used in this book

Geometrical   Surface area of a sphere $= 4\pi r^2$

Volume of a sphere $= 4\pi r^3/3$

Relativity $E = mc^2$

$$T_{(moving)} = T_{stationary}(1/(1-v^2))^{0.5}$$

where T is time and v (italics) is the velocity expressed as a fraction of c

$$Mass_{stationary} = Mass_{(moving)} (1/(1-v^2))^{0.5}$$

Doppler $\lambda_{(moving)} = \lambda_{(Stationary)}((1+v)(1-v))^{0.5}$

where λ is the wavelength and v is again a fraction of c

$$z = (\lambda_{moving} - \lambda_{Stationary}) / \lambda_{Stationary})$$

Gravitational $g = GM_1 M_2/r^2$

where g is the gravitational acceleration in metres/sec^2, G is the gravitational constant, M_1 and M_2 are the masses in kg of two bodies r metres apart.

$$g = mv^2/r$$

where v is the orbital velocity of a body of mass m that is r metres from the central star or sun which is exerting g the gravitational acceleration at the distance r metres

Hubble $\qquad H_O = 652/(Age-D_T)$

where H_o is kilometres per sec, per Mega parsec, Age is the Age of the universe in billions of years (earth time). D_T is the time for light to travel distance D.

$$Age = D_T(2 + 3v)/3v$$

Distance to a supernova $\qquad D = (M + m + k - r - 25)/5$

where D is the distance in megaparsecs (base ten logarithm units), M is a reference standard magnitude of a type 1a supernova and has a value of -19.7, m is the magnitude of the observed supernova at a particular wavelength, k is a standardised correction factor to cover all wavelengths, r is a relativity correction factor. The numbers correct for magnitude's logarithmic scale..

Pendulum $\qquad T = 2\pi(l/g)^{0.5}$

Where T is time for the period of the pendulum in seconds, l is the length of the pendulum in metres, g is the gravitational acceleration experienced at ground level on earth experessed as metres per sec per sec.

INDEX

Printed in the United Kingdom
by Lightning Source UK Ltd.
133188UK00002B/19-27/P